Communications in Computer and Information Science 1786

Rationale
The CCIS series is devoted to the publication of proceedings of computer science conferences. Its aim is to efficiently disseminate original research results in informatics in printed and electronic form. While the focus is on publication of peer-reviewed full papers presenting mature work, inclusion of reviewed short papers reporting on work in progress is welcome, too. Besides globally relevant meetings with internationally representative program committees guaranteeing a strict peer-reviewing and paper selection process, conferences run by societies or of high regional or national relevance are also considered for publication.

Topics
The topical scope of CCIS spans the entire spectrum of informatics ranging from foundational topics in the theory of computing to information and communications science and technology and a broad variety of interdisciplinary application fields.

Information for Volume Editors and Authors
Publication in CCIS is free of charge. No royalties are paid, however, we offer registered conference participants temporary free access to the online version of the conference proceedings on SpringerLink (http://link.springer.com) by means of an http referrer from the conference website and/or a number of complimentary printed copies, as specified in the official acceptance email of the event.

CCIS proceedings can be published in time for distribution at conferences or as post-proceedings, and delivered in the form of printed books and/or electronically as USBs and/or e-content licenses for accessing proceedings at SpringerLink. Furthermore, CCIS proceedings are included in the CCIS electronic book series hosted in the SpringerLink digital library at http://link.springer.com/bookseries/7899. Conferences publishing in CCIS are allowed to use Online Conference Service (OCS) for managing the whole proceedings lifecycle (from submission and reviewing to preparing for publication) free of charge.

Publication process
The language of publication is exclusively English. Authors publishing in CCIS have to sign the Springer CCIS copyright transfer form, however, they are free to use their material published in CCIS for substantially changed, more elaborate subsequent publications elsewhere. For the preparation of the camera-ready papers/files, authors have to strictly adhere to the Springer CCIS Authors' Instructions and are strongly encouraged to use the CCIS LaTeX style files or templates.

Abstracting/Indexing
CCIS is abstracted/indexed in DBLP, Google Scholar, EI-Compendex, Mathematical Reviews, SCImago, Scopus. CCIS volumes are also submitted for the inclusion in ISI Proceedings.

How to start
To start the evaluation of your proposal for inclusion in the CCIS series, please send an e-mail to ccis@springer.com.

Matthew Forshaw · Katja Gilly ·
William Knottenbelt · Nigel Thomas
Editors

Practical Applications of Stochastic Modelling

11th International Workshop, PASM 2022
Alicante, Spain, September 23, 2022
Revised Selected Papers

Editors
Matthew Forshaw
Newcastle University
Newcastle upon Tyne, UK

Katja Gilly
Universidad Miguel Hernández de Elche
Elche, Spain

William Knottenbelt
Imperial College London
London, UK

Nigel Thomas
Newcastle University
Newcastle upon Tyne, UK

ISSN 1865-0929 ISSN 1865-0937 (electronic)
Communications in Computer and Information Science
ISBN 978-3-031-44052-6 ISBN 978-3-031-44053-3 (eBook)
https://doi.org/10.1007/978-3-031-44053-3

This Springer imprint is published by the registered company Springer Nature Switzerland AG
The registered company address is: Gewerbestrasse 11, 6330 Cham, Switzerland

Paper in this product is recyclable.

Preface

This volume of CCIS contains papers presented at the Eleventh International Workshop on Practical Applications of Stochastic Modelling (PASM 2022) held in Santa Pola, Alicante, Spain in September 2022.

Although small in number, the accepted papers demonstrate a diverse set of applications and approaches. PASM was collocated with the 18th European Performance Engineering Workshop (EPEW). This was the first time that PASM and EPEW were held in person since the Covid-19 pandemic, and it was a delight to once again be able to meet and converse with the presenters and other attendees. While the overall number of papers and presentations was lower than in some previous events, it was nevertheless an important step in renormalising workshop participation and sharing high-quality work in a positive and productive environment. A total of 7 full papers were accepted for PASM after going through a single-blind revision process that guaranteed a minimum of 3 reviews for each paper.

As workshop co-chairs we would like to thank everyone involved in making PASM and EPEW 2022 a success: Springer for their support of the workshop series, the programme committee and reviewers, and of course the authors of the papers submitted, without whom there could not be a workshop. We would especially like to extend our thanks to the staff of the Museo del Mar at the Castillo Fortaleza, Santa Pola, for hosting the workshop and providing an excellent environment and support. We trust that you, the reader, find the papers in this volume interesting, useful and inspiring.

November 2022

Matthew Forshaw
Katja Gilly
William Knottenbelt
Nigel Thomas

Organization

General Chairs

Matthew Forshaw	Newcastle University, UK
Katja Gilly	Universidad Miguel Hernández de Elche, Spain
William Knottenbelt	Imperial College London, UK
Nigel Thomas	Newcastle University, UK

Program Committee

Andrea Marin	Università Ca' Foscari di Venezia, Italy
Jean-Michel Fourneau	Université de Versailles, France
Charles Morisset	Newcastle University, UK
Marco Gribaudo	Politecnico di Milano, Italy
Mauro Iacono	Università degli Studi della Campania "Luigi Vanvitelli", Italy
Carlos Juiz	Universitat de les Illes Balears, Spain
Joris Walraevens	Universiteit Gent, Belgium
Samuel Kounev	Julius-Maximilians-Universität Würzburg, Germany
Sabina Rossi	Università Ca' Foscari di Venezia, Italy
Enrico Vicario	Università degli Studi di Firenze, Italy
Andrea Vandin	Sant'Anna School for Advanced Studies, Italy
Markus Siegle	Universität der Bundeswehr München, Germany
Aad van Moorsel	University of Birmingham, UK
Miklos Telek	Budapest University of Technology and Economics, Hungary

Additional Reviewers

Salvador Alcaraz
Paul Ezhilchelvan
Alexander Gouberman
Illes Horvath
Pedro Roig
Katinka Wolter

Contents

Performance Modelling of Attack Graphs

Ohud Almutairi and Nigel Thomas(✉)

Newcastle University, Newcastle upon Tyne, UK
{o.m.m.almutairi2,nigel.thomas}@newcastle.ac.uk

Abstract. This paper represents an initial study into using Performance Evaluation Process Algebra (PEPA) to model and analyse attack graphs. Such an approach adds timing information into the model and therefore extends the range of available analysis techniques. Two methods are proposed to generate a PEPA model based on a pre-existing attack graph with known vulnerabilities. The first method builds a PEPA model consisting of a single sequential component representing both a system and an attacker. The second method generates two sequential components and the system equation. The created PEPA models allow us to perform path analysis, sensitivity analysis and to estimate the time it takes for an attacker to compromise a system. We present two case studies of building and evaluating PEPA models of an attack graph. The PEPA Eclipse plug-in is used to support the evaluation of the PEPA model. We perform passage-time analysis on the models for each attack path in the attack graph, from the first vulnerability in a path until the system was compromised by the attacker. The results illustrate the most and least threatening attack paths and the time it takes the attacker to compromise the system for each path in the attack graph. They also show the impact of the attacker skills and the probability of exploit code availability on an attacker's time to compromise the system.

Keywords: PEPA · attack graph · Performance modelling

1 Introduction

Defending a system and keeping it secure is not trivial. A defender must keep a system secure by preventing or early detecting an attacker's intent in order to respond and recover in good time. An attack graph is a popular graph-based method proposed by Swiler *et al.* [13]. It can support a defender to understand an attacker's behaviour and then work to protect a system [8]. Attack graphs present the different attack paths that an attacker can follow to exploit multiple vulnerabilities in a system in order to achieve its final goal [7,8]. In an attack graph, each attack path is a sequence of nodes representing the vulnerabilities or exploits in a system [8], where edges represent the transitions between different nodes due to the attacker's actions [13]. Doing a risk assessment based on the attack graph of a system allows a defender to see the system from an attacker's

M. Forshaw et al. (Eds.): PASM 2022, CCIS 1786, pp. 1–26, 2023.
https://doi.org/10.1007/978-3-031-44053-3_1

perspective. The defender can perform path analysis to identify the most significant threatening path that an attacker could take to breach a system [13] and sensitivity analysis to show how hardening the security of some nodes would affect the security status of the system.

Constructing and evaluating a stochastic process algebra model version of a pre-existing attack graph of a system can allow us to do dynamic analysis in order to understand attacker behaviour, identify critical threats and estimate the time to compromise a system. This information can aid in prioritising the implementation of countermeasures necessary to maintain the security of a system. The time to compromise a system is the time it takes the attacker to successfully compromise the system [10], which is critical in determining the amount of safe time, or reaction time, a system has before being compromised [10]. A number of factors can also be considered to indicate how fast an attacker can compromise a system, such as the different attackers' capabilities [3,9] and the availability of exploit code for a vulnerability [3]. The attackers are classified based on their capabilities into beginner, intermediate and expert [9]. The beginner attacker can attack a system by using a pre-existing exploit code, the intermediate attacker can use and/or modify a pre-existing exploit code and an expert attacker can use and modify a pre-existing exploit code or create a new exploit code [9].

This paper represents an initial attempt to explore the modelling and analysis of attack graphs using the stochastic process algebra PEPA. The paper presents two methods to create a PEPA model for a pre-constructed attack graph with its vulnerabilities that already assigned attack probability from the international standard CVSS. The first method is to create a PEPA model that comprises of system and attacker as a coupled component. The second method is to create a PEPA model that comprises separate components of system and attacker and an equation for the model with the cooperation set. Then, we show through case studies how we can use the created PEPA models to do path analysis, sensitivity analysis and estimate an attacker's time to compromise a system. Moreover, in the second case study, the time is taken until the attacker compromises the system is estimated for each path by using the vulnerabilities that already existed in the system and considering different factors such as exploit code availability and the attacker's skill.

The paper is organized as follows. Section 2 presents some existing related studies. Section 3 presents methods to create a PEPA model for a pre-constructed attack graph. In Sects. 4 and 5, the case studies of building and evaluating PEPA model of a pre-constructed attack graph are presented. Finally, Sect. 6 concludes the paper by providing an overview of the study findings and future work.

2 Related Work

Many studies have been conducted in the field of risk assessment using attack graphs. Zheng *et al.* [14] present a quantitative method based on an attack graph and the international standard CVSS (Common Vulnerability Scoring System) for assessing network security risk levels and quantifying node reachability and importance. The evaluation of their model can assist security professionals in providing the network with an appropriate countermeasure. Sun *et al.* [12] propose a Network Security Risk Assessment Model (NSRAM) based on attack graph and Markov chain to provide the optimal attack path. Their proposed model generates the attack graph, and then each vulnerability is assigned attack probability from the international standard CVSS. Then they use a discrete Markov chain method to calculate the transition probability from node to node. The enterprise network security risk level and the attack probability of each path are provided using their proposed model. Their analysis is based on counting the steps till the system is compromised. However, estimating the actual time to compromise the system is important to indicate how much safe time the system has before it is compromised.

Pokhrel and Tsokos in [11] propose a stochastic model using a host access attack graph to evaluate the overall security risk of a network. Their model is based on Markov chains and a host access attack graph, with each vulnerability assigned a CVSS score. They introduce a bias factor to model the attacker's skill. Their findings can assist network administrators in prioritising the patching of vulnerable nodes in the network. Kaluarachchi *et al.* [6] propose a stochastic model using a Markovian approach to predict the Expected Path Length, which is the number of attacker's step to reach the final goal. Also, they provide the minimum attacker's steps to reach the final goal. Their findings can be used to assist security administrators in clearly understanding the security status of their system.

Abraham and Nair [1] propose a non-homogenous Markov model using an attack graph to assess the security state of the network by considering a temporal factor associated with the vulnerabilities. They focus on their security analysis to estimate expected path length, probabilistic path and expected impact metrics. In addition, Abraham [2] proposes a stochastic model based on a non-homogenous continuous-time Markov model to estimate the overall mean time to compromise by considering the casual relationship between the vulnerabilities in the attack graph and the attacker skill. In their model, the coefficients of different attacker skills were estimated by analysing 15 years of vulnerability data from the National Vulnerability Database (NVD). His results show that when the attacker has a higher skill to compromise the system, that causes the time to compromise the system to reduce.

In contrast to previous studies, we propose methods for converting a pre-existing attack graphs to PEPA models. PEPA supports a compositional, formal and abstract approach to construct a model for an attack graph. The model represents an attacker behaviour and the interaction between an attacker and a system. PEPA models allow us to perform dynamic analysis. They can be analysed via a continuous-time Markov chain with rates to support estimating the time to compromise the system for each attack path and the time it takes the attacker to get to a particular vulnerability in a system.

3 Methods for Creating a PEPA Model for a Given Attack Graph

3.1 Basic PEPA Model

We propose an algorithm for constructing a PEPA model from a pre-constructed attack graph with known vulnerabilities. The PEPA model comprises of a coupled component of system and attacker. For each node in the attack graph, the vulnerability name is the action name. The action rate is based on the probability of the vulnerability being breached and the transition probability. The probability of the vulnerability being breached is defined by the complexity of the attack in CVSS. Failed action and rate for each node are introduced in the model. Failed action occurs when the attacker fails to exploit the vulnerability. Failed rate is the probability of an attacker failing to exploit a vulnerability. When the attacker fails to exploit the vulnerability, the attacker returns to the root node in the attack graph.

The algorithm's input is an attack graph specification, which contains information about the relationship between nodes in an attack graph, and the vulnerability name and the probability of the vulnerability being breached for each node. The transition probability from one node to the next node is calculated by Markov chain as in [12]. For example, Fig. 1 illustrates a simple example of an attack graph. The breach probability of node H1 is 0.4, and the breach probability of node H2 is 0.8. The transition probability from Start node to node H1 is 0.33, resulting from the following equation:

$$P_{SH1} = \frac{bH1}{bH1 + bH2}$$

where P_{SH1} is the transition probability from Start node to node H1, $bH1$ is the breach probability of H1 and $bH2$ is the breach probability of H2.

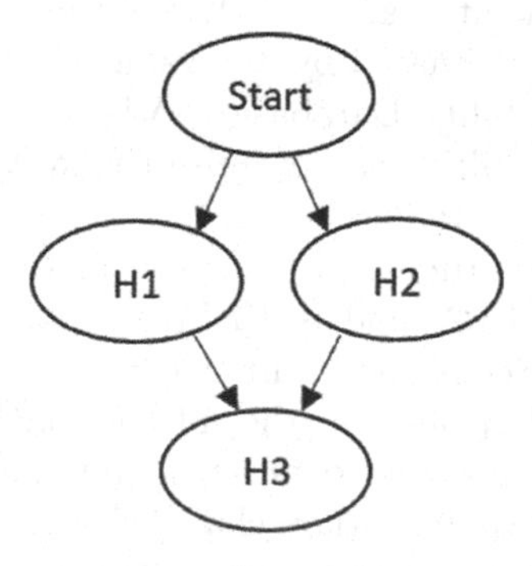

Fig. 1. An example of an attack graph.

Algorithm1. Create the system and attacker component based on a pre-existing attack graph

```
1: print "Start=(start,1).SystemAttacker"+first's node name";"
2: for all node ∈ attackGraph do
3:    if #directNeighbour_node > 0 & node ! = goalnode then
4:       print "SystemAttacker"+node name+"="
5:       s ← #directNeighbour_node
6:       for all neigh ∈ directNeighbour_node do
7:          s ← s − 1
8:          convert at least first letter of neigh's vulnerability name to lower case
9:          print "("+neigh's vulnerability name+","
            +node's BreachProbability*neigh's TransitionProbability
            +").SystemAttacker" +neigh's name
10:         if s = 0 then
11:            if this node ! = first node in attackGraph then
12:               print "(failed,"+ (1 − node's BreachProbability)+")
                  .System"+attackGraph's first node' name ";"
                  +new line
13:            else
14:               print ";"+new line
15:            end if
16:         else
17:            print "+"
18:         end if
19:      end for
20:   else
21:      print "SystemAttacker"+node's name+"=(compromised,"
         +node's BreachProbability +").Completed+(failed,"
         +(1 − node's BreachProbability)+").SystemAttacker"
         + attackGraph's first node' name";" + new line
22:      print "Completed=(completed,1).Start;"
23:   end if
24: end for
```

3.2 Multiple Component PEPA Model

This algorithm generates a PEPA model from a pre-constructed attack graph with known vulnerabilities. The PEPA model comprises separate of system and attacker components and an equation for the model with the cooperation set. As in the first algorithm, the vulnerability name is the action and the rate is based on the probability of the vulnerability being breached and the transition probability. We assumed that when the attacker fails to exploit the vulnerability, that attacker stays on the same node.

The inputs are:

- an attack graph: the relationship between nodes, the vulnerabilities names, and the probability of a vulnerability to be breached for each node. The transition probability between nodes is calculated same as in Subsect. 3.1.
- An attacker's abilities/steps that he takes to attack a system: the sequence of steps, the steps names, and rate.

Algorithm2. Create a system component based on the attack graph given

---

1: **print** "Start=(start,1).System"+first's *node* name";"
2: **for all** $node \in attackGraph$ **do**
3: **if** $\#directNeighbour_{node} > 0$ & $node$! = $goalnode$ **then**
4: **print** "System"+*node* name+"="
5: $s \leftarrow \#directNeighbour_{node}$
6: **for all** $neigh \in directNeighbour_{node}$ **do**
7: $s \leftarrow s - 1$
8: convert at least first letter of *neigh*'s vulnerability name to lower case
9: **print** "("+*neigh*'s vulnerability name+"," +*node*'s BreachProbability**neigh*'s Transition-Probability+").System" +*neigh*'s name
10: **if** $s = 0$ **then**
11: **if** this *node* ! = first *node* in *attackGraph* **then**
12: **print** "(failed,"+ (1 − *node*'s BreachProbability)+") .System"+*node*' name+";"+new line
13: **else**
14: **print** ";"+new line
15: **end if**
16: **else**
17: **print** "+"
18: **end if**
19: **end for**
20: **else**
21: **print** "System"+node's name+"=(compromised, " +*node*'s BreachProbability +").Completed+(failed," +(1 − *node*'s BreachProbability)+").System" +*node*' name";" + new line
22: **print** "Completed=(completed,1).Start;"
23: **end if**
24: **end for**

Algorithm3. Create an attacker component based on the given attacker's steps to attack a system and the attack graph

---

1: convert at least first letter of attacker's name to upper case
2: **print** *attacker*'s name+"Start=(start,1)."+ *attacker*'s name+ first *node*'s name+";"

```
3: for all node ∈ attackGraph do
4:   if node ! = first node in attackGraph then
5:     for all step ∈ attackerSteps do
6:       print attacker's name+step's number +node's name
          +"="
7:       if #directNextSteps_step > 0 then
8:         r ← #directNextSteps_step
9:         for all nextStep ∈ directNextSteps_step do
10:          r ← r − 1
11:          convert at least first letter of step's name to lower case
12:          print "("+ step's name+","+ step's rate +")."
             +attacker's name+ nextStep's number
             +node's name
13:          if r = 0 then
14:            print ";" + new line
15:          else
16:            print "+"
17:          end if
18:        end for
19:      else
20:        convert at least first letter of step's name to lower case
21:        print "("+ step's name +","+ step's rate+")."
           +attacker's name + node's name +";" + new line
22:      end if
23:    end for
24:  end if
25:  if #directNeighbour_node > 0 & node ! = goalnode then
26:    print attacker's name + node's name +"="
27:    s ← #directNeighbour_node
28:    for all neigh ∈ directNeighbour_node do
29:      s ← s − 1
30:      convert at least first letter of neigh's vulnerability name to lower case

31:      print "("+neigh's vulnerability name+","
         +node's BreachProbability*neigh's TransitionProbability
         +")." +attacker's name+attacker's first step's number
         +neigh's name
32:      if s = 0 then
33:        if this node ! = first node in attackGraph then
34:          print "(failed,"+ (1 − node's BreachProbability)+")."
             +attacker's name+first step's number +node's
              name+";"+new line
35:        else
36:          print ";"+new line
37:        end if
```

```
38:         else
39:            print "+"
40:         end if
41:      end for
42:   else
43:      print attacker's name+node's name+"=(compromised, "
         +node's BreachProbability +").Completed+(failed,"
         +(1 − node's BreachProbability)+")."
         +attacker's name+first step's number +node's name+";"
44:      print "Completed=(completed,1)."+ attacker's name
         +"Start; "
45:   end if
46: end for
```

Algorithm4. Create a system equation for the model

```
 1: print attacker's name+ "Start<start, failed, compromised,
    completed"
 2: for all node ∈ attackGraph do
 3:    if node's vulenrabilty's name ! = null then
 4:       if node ! = last node in attackGraph then
 5:          print node's vulnerability' name+","
 6:       else
 7:          print node's vulnerability' name
 8:       end if
 9:    end if
10: end for
11: print ">"+"Start"
```

4 Case Study 1: Basic PEPA Model for Attack Graph

4.1 System Specification

The attack graph is taken from Sun *et al.* [12].

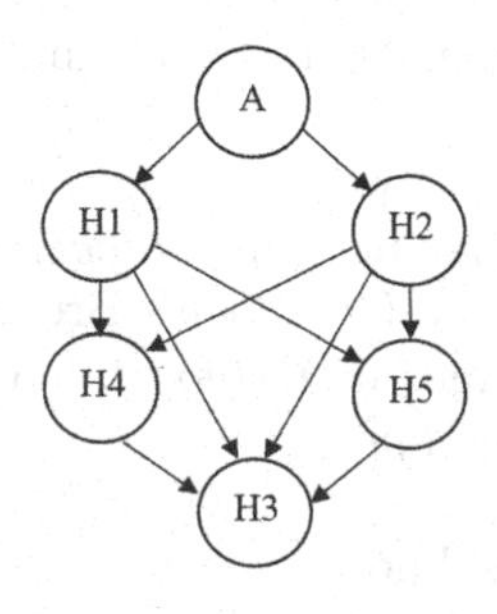

Fig. 2. Attack graph [12].

As in [12], the attack graph comprises of six nodes, Fig. 2. Node A is the starting node, and node H3 is the target node. H1, H2, H3, H4 and H5 are hosts in the system that an attacker tries to exploit their vulnerabilities in order to reach the final goal. Each host has a vulnerability name and a probability of being exploited as shown in Table 1.

Table 1. The probabilities of vulnerabilities to be breached [12].

Host	H1	H2	H3	H4	H5
The vulnerability	servU5	telnet	sql	rpc	remoteLog
The breach probability	0.4	0.8	0.6	0.5	0.9

4.2 PEPA Model

The following is a valid PEPA model resulted from implementing and running Algorithm 1. In the PEPA model, there is one sequential component. This component is a coupled component representing the system and attacker moving sequentially from the different behaviours based on the activities specified in the model. The model is formulated as follows:

The Sequential Component of an Attacker and a System.

$$\begin{aligned}
Start &= (start, 1).SystemAttackerA;\\
SystemAttackerA &= (servU5, 0.33).SystemAttackerH1\\
&\quad + (telnet, 0.67).SystemAttackerH2;\\
SystemAttackerH1 &= (sql, 0.12).SystemAttackerH3\\
&\quad + (rpc, 0.1).SystemAttackerH4\\
&\quad + (remoteLogin, 0.18).SystemAttackerH5\\
&\quad + (failed, 0.6).SystemAttackerA;
\end{aligned}$$

$$\begin{aligned}
SystemAttackerH2 &= (sql, 0.24).SystemAttackerH3\\
&\quad + (rpc, 0.2).SystemAttackerH4\\
&\quad + (remoteLogin, 0.36).SystemAttackerH5\\
&\quad + (failed, 0.2).SystemAttackerA;\\
SystemAttackerH3 &= (compromised, 0.6).Completed\\
&\quad + (failed, 0.4).SystemAttackerA;\\
Completed &= (completed, 1).Start;\\
SystemAttackerH4 &= (sql, 0.5).SystemAttackerH3\\
&\quad + (failed, 0.5).SystemAttackerA;\\
SystemAttackerH5 &= (sql, 0.9).SystemAttackerH3\\
&\quad + (failed, 0.1).SystemAttackerA;
\end{aligned}$$

The first state in our model is *Start*. It is a starting point for an attacker to launch an attack on the system in our model. The attacker performs action *start* at rate 1 leading to state *SyatemAttackerA*, which represents the first node in the attack graph. Then, he can perform either action *servU*5 at rate 0.33, which is the transition probability to H1, leading to *SystemAttackerH*1 or *telnet* action at rate 0.67, which is the transition probability to H2, leading to *SystemAttackerH*2. *SystemAttackerH*1 and *SystemAttackerH*2 represent H1 node and H2 node in the attack graph, respectively.

When the attacker in the state *SystemAttackerH*1, there is one of four actions could happen either *sql* at rate 0.12, which is the product of the probability of *servU*5 being breached and the transition probability to H3, leading to *SystemAttackerH*3, *rpc* at rate 0.1 ,which is the product of the probability of *servU*5 being breached and the transition probability to H4, leading to *SystemAttackerH*4, *remoteLogin* at rate 0.18, which is the product of the probability of *servU*5 being breached and the transition probability to H5, leading to *SystemAttackerH*5 or *failed* at rate 0.6, which is the result of (1-the probability of *servU*5 being breached) ,leading to *SystemAttackerA*. *SystemAttackerH*3, *SystemAttackerH*4, *SystemAttackerH*5 and *SystemAttackerA* represent H3, H4, H5 and A node in the attack graph, respectively.

In the state *SystemAttackerH*2, there is also one of four actions that could be performed either *sql* at rate 0.24, which is the product of the probability of *telnet* being breached and the transition probability to H3, leading to *SystemAttackerH*3, *rpc* at rate 0.2, which is the product of the probability of *telnet* being breached and the transition probability to H4, leading to *SystemAttackerH*4, *remoteLogin* at rate 0.36, which is the product of the probability of *telnet* being breached and the transition probability to H5, leading to *SystemAttackerH*5 or *failed* at rate 0.2, which is the result of (1-the probability of *telnet* to be breached), leading to *SystemAttackerA*.

When the attacker in the state *SystemAttackerH*4, he can perform either *sql* at rate 0.5, which is the product of the probability of *rpc* being breached and the transition probability to H3, leading to *SystemAttackerH*3 or *failed* at rate 0.5, which is the result of (1- the probability of *rpc* to be breached), leading back to *SystemAttackerA*. In state *SystemAttackerH*5, one of two actions can happen either *sql* at rate 0.9, which is the product of the probability of *remoteLogin* being breached and the transition probability to H3, leading to *SystemAttackerH*3 or *failed* at rate 0.1, which is the result of (1-the probability of *remoteLog* to be breached), leading back to *SystemAttackerA*.

In the state *SystemAttackerH*3 which represents the attacker starts to attack the target node by performing either action *compromised* at rate 0.6, which is the probability of *sql* being breached, leading to *Completed* state or action *failed* at rate 0.4, which is the result of (1-the probability of *sql* to be breached) leading back to *SystemAttackerA*. *Completed* is the state that represents the system is successfully compromised. In the *Completed* state, the attacker performs action *complete* at rate 1 returning back to the *Start* state so that the model becomes cyclic and steady-state measures can be obtained.

4.3 Performance Evaluation

Path Analysis. There are six possible paths that the attacker can use to compromise the system [12]. We are interested in estimating the time to compromise, which is the amount of time it takes the attacker to compromise the system. Note that the time units in these examples are arbitrary, as we lack data on the speed at which the attack progresses. Therefore, we are interested in exploring the relative time to compromise via different paths. The time to compromise is calculated for each path in the attack graph. We perform passage-time analysis for each path. In this model, when an attacker fails to exploit a node's vulnerability, the attacker returns to the attack graph's first node.

Figure 3 shows the passage-time for each path in the attack graph from the first vulnerability action to *compromised* action. If the attacker fails to exploit the vulnerability, he returns to *SystemAttackerA* state which is the root node in the attack graph.

Table 2 is taken from Sun *et al.* [12]. It shows the attack path and the attack probability of each path. Path 1 has the lowest attack probability, whereas path 6 has the highest attack probability. Figure 3 shows our evaluation result which is the time until the system is compromised for each path in the attack graph.

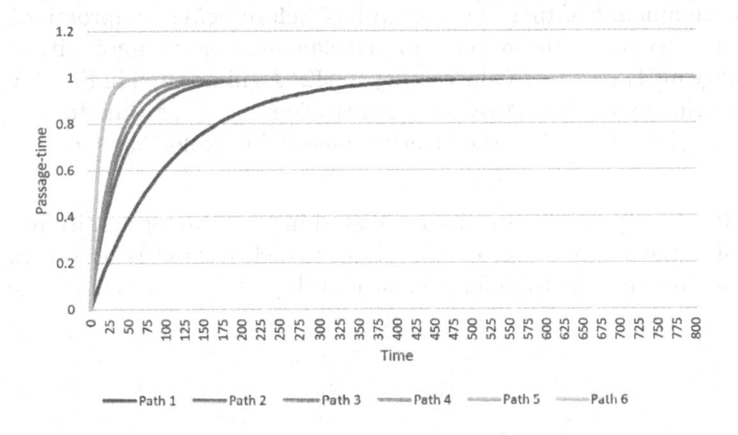

Fig. 3. Passage-time analysis for each path in the attack graph.

Table 2. Attack Paths and probabilities, Sun *et al.* [12].

Path number	Starting node	Target node	The path	Attack probability
1	A	H3	A→H1→H4→H3	0.0099
2	A	H3	A→H1→H3	0.02376
3	A	H3	A→H1→H5→H3	0.032076
4	A	H3	A→H2→H4→H3	0.0402
5	A	H3	A→H2→H3	0.09684
6	A	H3	A→H2→H5→H3	0.130248

Table 3. The path order from our result and Sun *et al.*.

The fastest path form PEPA model	The highest attack probability from [12]
Path 6	Path 6
Path 5	Path 5
Path 4	Path 4
Path 3	Path 3
Path 2	Path 2
Path 1	Path 1
The slowest path	The lowest attack probability

The fastest path is path 6, which has the highest attack probability, as shown in Table 2. The slowest path is path 1, which has the lowest attack probability, as shown in Table 2.

Table 3 presents the fastest to slowest paths from our PEPA model evaluation result and the highest to lowest attack probabilities from Sun *et al.* Our evaluation result has the same path order as Sun *et al.* It shows clearly the most threatening path to the lowest threatening path in the attack graph. Figure 3 indicates the amount of time the system has before being compromised.

Moreover, to attack the system, an attacker must first exploit either *servU5* vulnerability in H1 or *telnet* vulnerability in H2. As illustrated in Fig. 4, when the attacker begins by exploiting *telnet*, the attacker's average time to compromise the system is less than when the attacker begins by exploiting *servU5*.

Sensitivity Analysis. We are also interested in undertaking sensitivity analysis to determine the sensitivity of each path an attacker could take to compromise the system. We are going to change the probability of a *telnet* being breached and

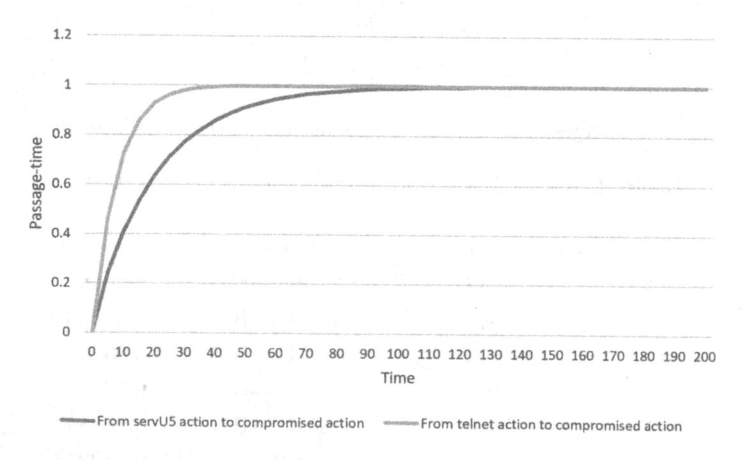

Fig. 4. Passage-time analysis for exploiting *servU5* and *telnet* to *compromised* action.

then perform passage-time analysis on each path to determine how that affects the time to compromise for each path. We change the *telnet* action rate to 0.5 and then to 0.1 and keep all other actions unchanged. This makes breaching this vulnerability much harder. Figures 5 and 6 show when we reduce the probability of *telnet* to be breached, the time takes the attacker to successfully reach its final goal increases for path 6, 5 and 4.

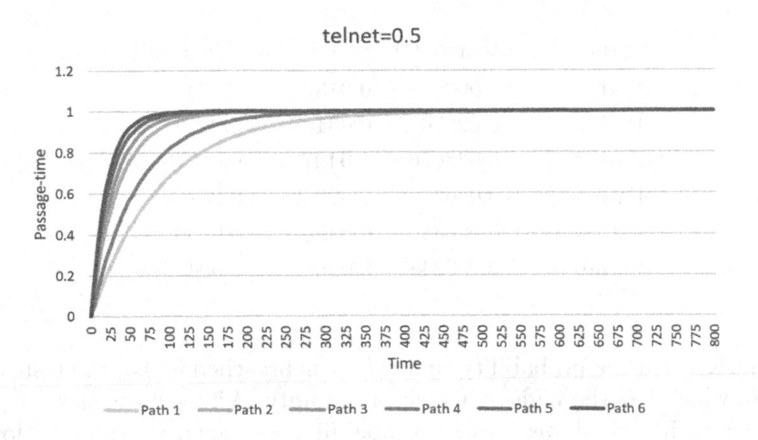

Fig. 5. Passage-time analysis for all paths when *telnet* action rate=0.5.

Following approach of Sun *et al.* [12] to calculate the attack probability of each path, the new attack probability after changing the value of *telnet* is presented in Table 4. In Table 4, the attack probability of each path when the probability of *telnet* to be breached is 0.8 is taken from [12]. The attack probability of each path when the probability of *telnet* to be breached is changed to 0.5 and 0.1 are calculated by following the same approach that Sun *et al.* follow. As you

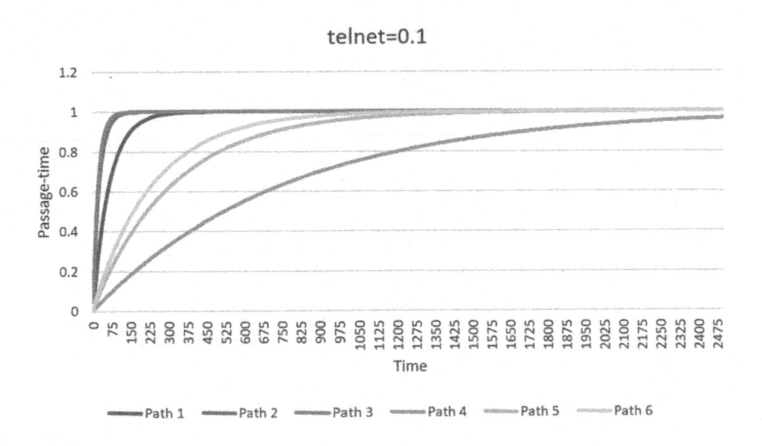

Fig. 6. Passage-time analysis for all paths when *telnet* action rate=0.1.

can see from Table 4, there is a significant decrease in the attack probability of path 4, 5 and 6 when the probability of *telnet* to be breached is reduced. This is shown in our evaluation results, Figs. 5 and 6, as an increase in the attacker's time to compromise. The attack probability of path 1, 2 and 3 slightly increased.

Table 4. The attack probability of each path when the probability of *telnet* to be breach is equal to 0.8, 0.5 and 0.1.

Attack path	telnet=0.8	telnet=0.5	telnet=0.1
Path 1	0.0099	0.0132	0.024
Path 2	0.02376	0.03168	0.0576
Path 3	0.032076	0.042768	0.07776
Path 4	0.0402	0.020625	0.0015
Path 5	0.09684	0.0495	0.0036
Path 6	0.130248	0.066825	0.00486

In Fig. 5, when the probability that *telnet* is breached is 0.5, the fastest path is path 6, which has the highest attack probability. Whereas the slowest path is path 1, which has the lowest attack probability as shown in Table 4. However, path 6 in Fig. 3 when the probability that *telnet* is breached is 0.8 is faster than path 6 in Fig. 5. When the attacker follows path 6 to compromise the system, the attacker's time to compromise is 155 time units when the probability of *telnet* to be breached is 0.8 and 290 time units when the probability of *telnet* to be breached is 0.5. In Fig. 6, when the probability of *telnet* to be breached is reduced to 0.1, the fastest path becomes path 3, which has the highest attack probability. Whereas the slowest path is path 4, which has the lowest attack probability as it is shown in Table 4. That shows that our evaluation result reflects the change in the probability of *telnet* to be breached.

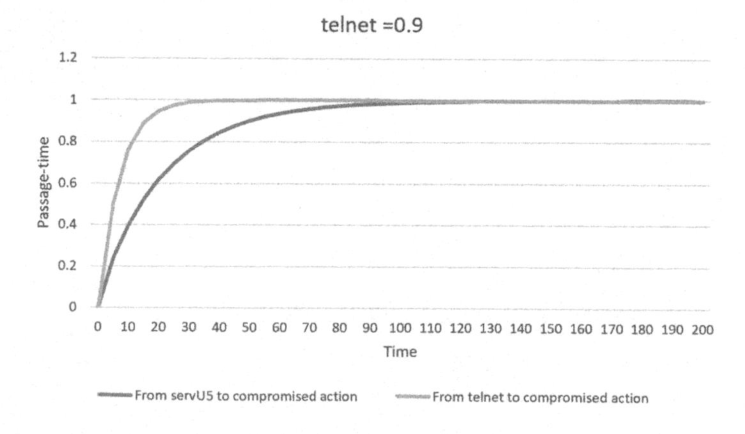

Fig. 7. Passage-time analysis when *telnet* action rate =0.9.

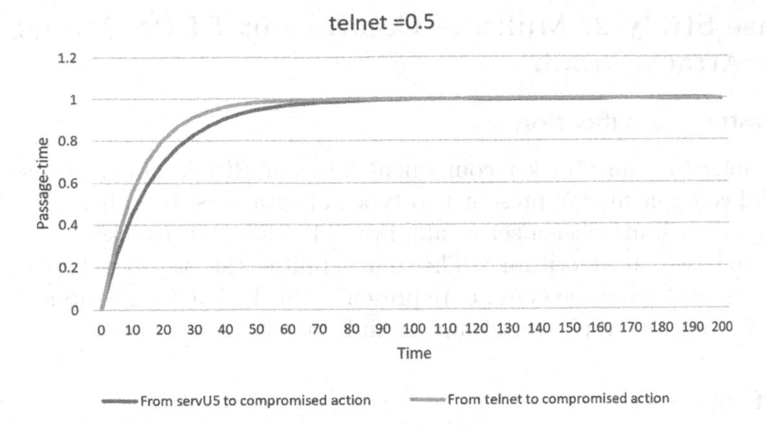

Fig. 8. Passage-time analysis when *telnet* action rate =0.5.

To compromise the system, the attacker must first attack H1 via *servU*5 or H2 via *telnet*. We adjust the *telnet* action rate to 0.9, 0.5, and 0.1 while maintaining all other actions unchanged to determine the effect on the attacker's time to compromise. Changing the probability of *telnet* to be breached could significantly impact the attacker's time to compromise the system. The probability could be reduced by implementing some protecting tools or security measures in host H2 to make it difficult for the attacker to exploit the vulnerability. Figures 7, 8 and 9 illustrate how reducing the probability rate of *telnet* action to be breached causes the attacker's time to compromise the system to increase when the attacker attacks the system by exploiting *telnet*.

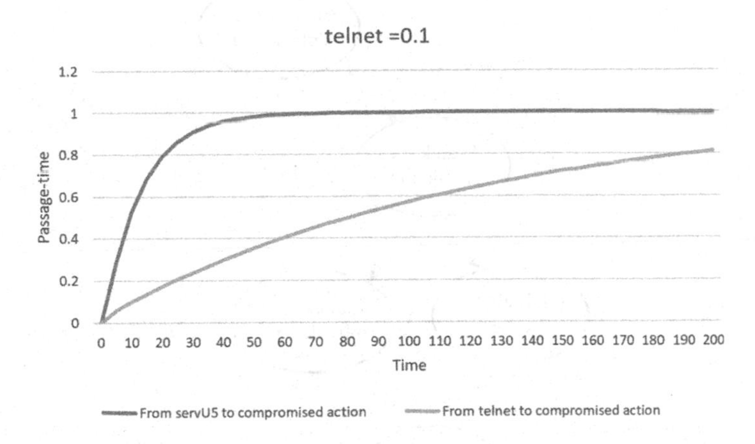

Fig. 9. Passage-time analysis when *telnet* action rate =0.1.

5 Case Study 2: Multiple Component PEPA Model for Attack Graph

5.1 System Specification

We now introduce an attacker component into our PEPA model as a separate sequential component. We present two types of attackers: beginner and expert. We assigned each attacker a set of abilities. The different attacker components are created based on Algorithm 3. The steps/abilities that the attackers follow to support compromising the system are presented in the following graphs, Figs. 10 and 11. The attack graph is the same as in Subsect. 4.1.

5.2 PEPA Model

The following is a valid PEPA model resulted from implementing and running Algorithm 2 for system component, Algorithm 3 for attacker components and Algorithm 4 for system equation. In our PEPA model, there are two types of components: system and attacker. There are also two types of attacker components: beginner and expert. The components move sequentially from their different behaviours based on the activities specified in the model.

Moreover, to introduce an exploit code availability factor in our model, we introduce a rate of p, which represents the probability of an exploit code availability. The rate of *searchCode* actions in all attacker's states is modelled by having a probabilistic choice p to go to the state *Attacker*2*H*1 when the suitable code is found or to the state *Attacker*3*H*1 when the suitable exploit code is not found. The model is formulated as follows:

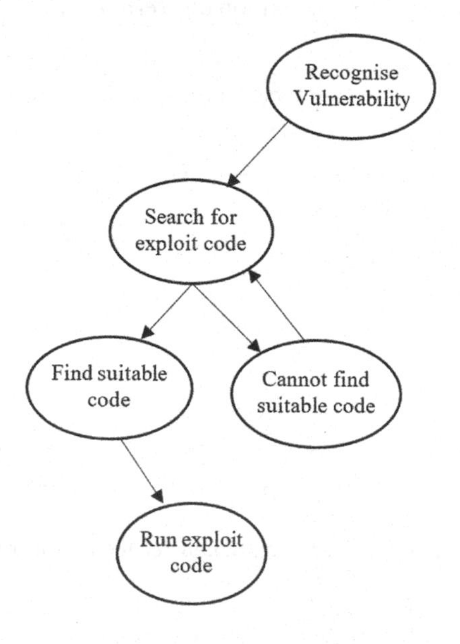

Fig. 10. The beginner attacker.

A System Component

$$\begin{aligned}
Start &= (start, 1).SystemA;\\
SystemA &= (servU5, 0.33).SystemH1\\
&\quad + (telnet, 0.67).SystemH2;\\
SystemH1 &= (sql, 0.12).SystemH3 + (rpc, 0.1).SystemH4\\
&\quad + (remoteLogin, 0.18).SystemH5\\
&\quad + (failed, 0.6).SystemH1;\\
SystemH2 &= (sql, 0.24).SystemH3 + (rpc, 0.2).SystemH4\\
&\quad + (remoteLogin, 0.36).SystemH5\\
&\quad + (failed, 0.2).SystemH2;\\
SystemH3 &= (compromised, 0.6).Completed\\
&\quad + (failed, 0.4).SystemH3;\\
Completed &= (completed, 1).Start;\\
SystemH4 &= (sql, 0.5).SystemH3 + (failed, 0.5).SystemH4;\\
SystemH5 &= (sql, 0.9).SystemH3 + (failed, 0.1).SystemH5;
\end{aligned}$$

The above model component specifies the System's different behaviours, moving from *Start* to *Completed*. The description of this component is similar to the component in Subsect. 4.2 except when a failed action happens the system stays in the same state it was in.

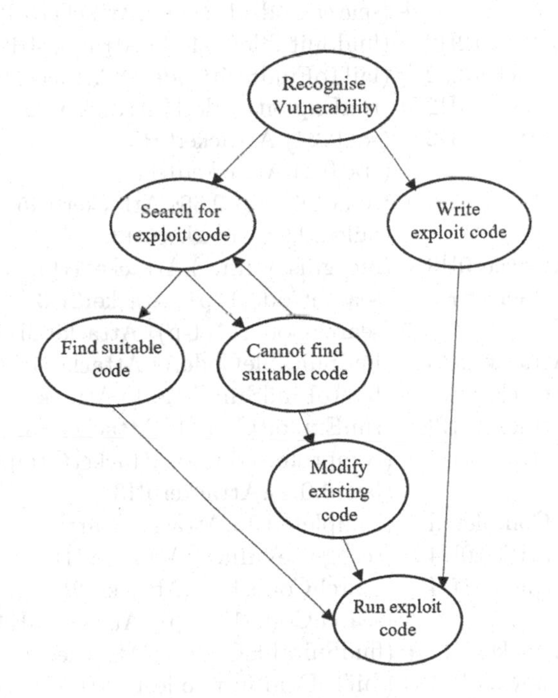

Fig. 11. The expert attacker.

The Beginner Attacker Component. This part of the model represents the beginner attacker's different behaviours, moving from *AttackerStart* to *AttackerCompleted* to compromise the system. The beginner attacker is just able to exploit the known vulnerability by using a pre-existing exploit code, as shown in Fig. 10.

$$
\begin{aligned}
\text{AttackerStart} &= (\text{start},1).\text{AttackerA};\\
\text{AttackerA} &= (\text{servU5},0.33).\text{Attacker0H1}\\
&\quad + (\text{telnet},0.67).\text{Attacker0H2};\\
\text{Attacker0H1} &= (\text{recogniseVuln},1).\text{Attacker1H1};\\
\text{Attacker1H1} &= (\text{searchCode},1*\text{p}).\text{Attacker2H1}\\
&\quad + (\text{searchCode},1*(1\text{-p})).\text{Attacker3H1};\\
\text{Attacker2H1} &= (\text{findSuitableCode},1).\text{Attacker4H1};\\
\text{Attacker3H1} &= (\text{failToFindSuitCode},1).\text{Attacker1H1};\\
\text{Attacker4H1} &= (\text{runExploitCode},1).\text{AttackerH1};\\
\text{AttackerH1} &= (\text{sql},0.12).\text{Attacker0H3}\\
&\quad + (\text{rpc},0.1).\text{Attacker0H4}\\
&\quad + (\text{remoteLogin},0.18).\text{Attacker0H5}\\
&\quad + (\text{failed},0.6).\text{Attacker0H1};\\
\text{Attacker0H2} &= (\text{recogniseVuln},1).\text{Attacker1H2};\\
\text{Attacker1H2} &= (\text{searchCode},1*\text{p}).\text{Attacker2H2}\\
&\quad + (\text{searchCode},1*(1\text{-p})).\text{Attacker3H2};\\
\text{Attacker2H2} &= (\text{findSuitableCode},1).\text{Attacker4H2};\\
\text{Attacker3H2} &= (\text{failToFindSuitCode},1).\text{Attacker1H2};\\
\text{Attacker4H2} &= (\text{runExploitCode},1).\text{AttackerH2};\\
\text{AttackerH2} &= (\text{sql},0.24).\text{Attacker0H3}\\
&\quad + (\text{rpc},0.2).\text{Attacker0H4}\\
&\quad + (\text{remoteLogin},0.36).\text{Attacker0H5}\\
&\quad + (\text{failed},0.2).\text{Attacker0H2};\\
\text{Attacker0H3} &= (\text{recogniseVuln},1).\text{Attacker1H3};\\
\text{Attacker1H3} &= (\text{searchCode},1*\text{p}).\text{Attacker2H3}\\
&\quad + (\text{searchCode},1*(1\text{-p})).\text{Attacker3H3};\\
\text{Attacker2H3} &= (\text{findSuitableCode},1).\text{Attacker4H3};\\
\text{Attacker3H3} &= (\text{failToFindSuitCode},1).\text{Attacker1H3};\\
\text{Attacker4H3} &= (\text{runExploitCode},1).\text{AttackerH3};\\
\text{AttackerH3} &= (\text{compromised},0.6).\text{AttackerCompleted}\\
&\quad + (\text{failed},0.4).\text{Attacker0H3};\\
\text{AttackerCompleted} &= (\text{completed},1).\text{AttackerStart};\\
\text{Attacker0H4} &= (\text{recogniseVuln},1).\text{Attacker1H4};\\
\text{Attacker1H4} &= (\text{searchCode},1*\text{p}).\text{Attacker2H4} +\\
&\quad + (\text{searchCode},1*(1\text{-p})).\text{Attacker3H4};\\
\text{Attacker2H4} &= (\text{findSuitableCode},1).\text{Attacker4H4};\\
\text{Attacker3H4} &= (\text{failToFindSuitCode},1).\text{Attacker1H4};\\
\text{Attacker4H4} &= (\text{runExploitCode},1).\text{AttackerH4};\\
\text{AttackerH4} &= (\text{sql},0.5).\text{Attacker0H3}\\
&\quad + (\text{failed},0.5).\text{Attacker0H4};
\end{aligned}
$$

```
Attacker0H5 = (recogniseVuln,1).Attacker1H5;
Attacker1H5 = (searchCode,1*p).Attacker2H5
            + (searchCode,1*(1-p)).Attacker3H5;
Attacker2H5 = (findSuitableCode,1).Attacker4H5;
Attacker3H5 = (failToFindSuitCode,1).Attacker1H5;
Attacker4H5 = (runExploitCode,1).AttackerH5;
 AttackerH5 = (sql,0.9).Attacker0H3
            + (failed,0.1).Attacker0H5;
```

The Expert Attacker Component. This part of the model represents the expert attacker's different behaviours, moving from *AttackerStart* to *AttackerCompleted* to compromise the system. The expert attacker can use and/or modify a pre-existing exploit code and write a new exploit code, as shown in Fig. 11.

```
AttackerStart = (start,1).AttackerA;
    AttackerA = (servU5,0.33).Attacker0H1
              + (telnet,0.67).Attacker0H2;
  Attacker0H1 = (recogniseVuln,1).Attacker1H1
              + (recogniseVuln,1).Attacker6H1;
  Attacker1H1 = (searchCode,1*p).Attacker2H1
              + (searchCode,1*(1-p)).Attacker3H1;
  Attacker2H1 = (findSuitableCode,1).Attacker4H1;
  Attacker3H1 = (failToFindSuitCode,1).Attacker1H1
              + (failToFindSuitCode,1).Attacker5H1;
  Attacker4H1 = (runExploitCode,1).AttackerH1;
  Attacker5H1 = (modifyExploitCode,1).Attacker4H1;
  Attacker6H1 = (writeExploitCode,1).Attacker4H1;
   AttackerH1 = (sql,0.12).Attacker0H3
              + (rpc,0.1).Attacker0H4
              + (remoteLogin,0.18).Attacker0H5
              + (failed,0.6).Attacker0H1;
  Attacker0H2 = (recogniseVuln,1).Attacker1H2
              + (recogniseVuln,1).Attacker6H2;
  Attacker1H2 = (searchCode,1*p).Attacker2H2
              + (searchCode,1*(1-p)).Attacker3H2;
  Attacker2H2 = (findSuitableCode,1).Attacker4H2;
  Attacker3H2 = (failToFindSuitCode,1).Attacker1H2
              + (failToFindSuitCode,1).Attacker5H2;
  Attacker4H2 = (runExploitCode,1).AttackerH2;
  Attacker5H2 = (modifyExploitCode,1).Attacker4H2;
  Attacker6H2 = (writeExploitCode,1).Attacker4H2;
   AttackerH2 = (sql,0.24).Attacker0H3
              + (rpc,0.2).Attacker0H4
```

+ (remoteLogin,0.36).Attacker0H5
+ (failed,0.2).Attacker0H2;
Attacker0H3 = (recogniseVuln,1).Attacker1H3
+ (recogniseVuln,1).Attacker6H3;
Attacker1H3 = (searchCode,1*p).Attacker2H3
+ (searchCode,1*(1-p)).Attacker3H3;
Attacker2H3 = (findSuitableCode,1).Attacker4H3;
Attacker3H3 = (failToFindSuitCode,1).Attacker1H3
+ (failToFindSuitCode,1).Attacker5H3;
Attacker4H3 = (runExploitCode,1).AttackerH3;
Attacker5H3 = (modifyExploitCode,1).Attacker4H3;
Attacker6H3 = (writeExploitCode,1).Attacker4H3;
AttackerH3 = (compromised,0.6).AttackerCompleted
+ (failed,0.4).Attacker0H3;
AttackerCompleted = (completed,1).AttackerStart;
Attacker0H4 = (recogniseVuln,1).Attacker1H4
+ (recogniseVuln,1).Attacker6H4;
Attacker1H4 = (searchCode,1*p).Attacker2H4
+ (searchCode,1*(1-p)).Attacker3H4;
Attacker2H4 = (findSuitableCode,1).Attacker4H4;
Attacker3H4 = (failToFindSuitCode,1).Attacker1H4
+ (failToFindSuitCode,1).Attacker5H4;
Attacker4H4 = (runExploitCode,1).AttackerH4;
Attacker5H4 = (modifyExploitCode,1).Attacker4H4;
Attacker6H4 = (writeExploitCode,1).Attacker4H4;
AttackerH4 = (sql,0.5).Attacker0H3
+ (failed,0.5).Attacker0H4;
Attacker0H5 = (recogniseVuln,1).Attacker1H5
+ (recogniseVuln,1).Attacker6H5;
Attacker1H5 = (searchCode,1*p).Attacker2H5
+ (searchCode,1*(1-p)).Attacker3H5;
Attacker2H5 = (findSuitableCode,1).Attacker4H5;
Attacker3H5 = (failToFindSuitCode,1).Attacker1H5
+ (failToFindSuitCode,1).Attacker5H5;
Attacker4H5 = (runExploitCode,1).AttackerH5;
Attacker5H5 = (modifyExploitCode,1).Attacker4H5;
Attacker6H5 = (writeExploitCode,1).Attacker4H5;
AttackerH5 = (sql,0.9).Attacker0H3
+ (failed,0.1).Attacker0H5;

System Equation. The system equation and complete specification are given by

$$AttackerStart < start, failed, compromised, completed, servU5, telnet, sql, rpc, remoteLogin > Start$$

The two components are initially in state *AttackerStart* and *Start*. The actions between them are shared actions between the two components.

5.3 Performance Evaluation of the PEPA Models

We are also interested in estimating the time required to compromise the system for each possible path taken by each attacker. We evaluate the PEPA models by performing passage-time analysis. The passage-time is calculated from the first vulnerability action in a path to *completed* action. If an attacker fails to exploit the vulnerability, it returns to the first step in the attacker's ability/steps graph to try again to attack the host.

We set the probability of exploit code availability to 0.2 ($p = 0.2$) to also show the impact of the lack of exploit code on each attacker. The time to compromise for each attack path in the attack graph for the different attackers when the probability of exploit code availability is 0.2 is shown in Figs. 12 and 13. The time to compromise of paths 1, 2, 3, 4, 5 and 6 for the expert attacker are less than the time to compromise of paths 1, 2, 3, 4, 5 and 6 for beginner attacker, respectively. The fastest path in our PEPA model for the attack graph is path 6 which has the highest attack probability in [12]. The time it takes the beginner attacker to compromise the system for path 6 is 430 time units, which is longer than the time it takes the expert attacker to compromise the system for path 6 which is 150 time units.

Figures 12 and 13 clearly illustrate the most threatening attack path, the least threatening attack path and the time takes the attacker to compromise the system for each path in the attack graph for each attacker. The time to compromise for each attack path can rank the risk of all attack paths. Moreover, the figures clearly show the significant effect of lack of exploit code on the beginner attacker. The time to compromise for each path is larger for beginner attacker compared to each attack path for the expert attacker.

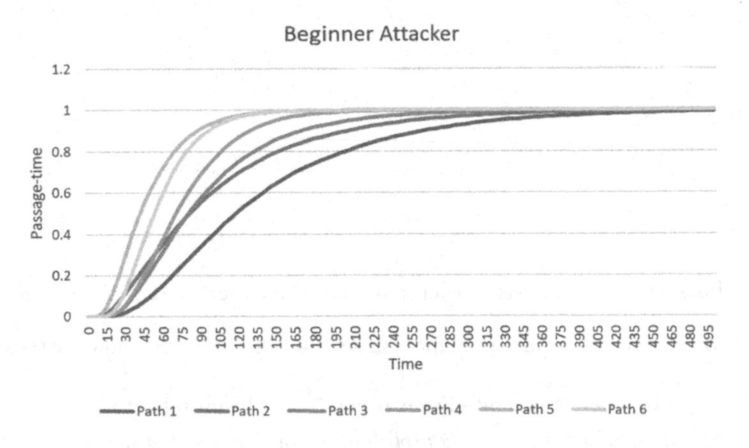

Fig. 12. Passage-time analysis of each path for the beginner attacker when $p = 0.2$.

Then, we change the probability of exploit code availability p from 0.2 to 0.8. Figures 14 and 15 illustrate the impact of the exploit code availability on the attacker's time to compromise of path 6 for each attacker. Path 6 is the fastest and most threatening path in our PEPA model for the attack graph. The impact of the availability of exploit code on the beginner attacker's time to compromise is significant as this attacker is only capable of using a pre-existing exploit code, as shown in Fig. 14. Whereas the expert attacker has a minimal impact as this attacker can modify and create exploit code when the suitable exploit code is not available, as shown in Fig. 15.

Moreover, as shown in the attack graph and our PEPA model, the attacker first needs to exploit either *servU*5 vulnerability in H1 or *telnet* vulnerability in H2 to start to attack the system. Figures 16 and 17 show when the beginner and expert attackers choose to exploit *telnet* first, the attacker's time to compromise the system is less than the time when the attacker starts to exploit *servU*5. This is because the probability of *telnet* being breached is 0.8, whereas the probability of *servU*5 to be breached is 0.4. It is much harder to exploit *servU*5 than *telnet*. However, the figures show that the expert attacker is faster than beginner attacker. In Figs. 16 and 17, the probability of exploit code availability for all vulnerabilities set to 0.2. Then, we set the probability of exploit code availability for all vulnerabilities to 0.8. Table 5 shows the significant impact of increasing the probability of exploit code availability for all vulnerabilities from 0.2 ($p = 0.2$) to 0.8 ($p = 0.8$) on the time to compromise the system for the beginner attacker. Whereas the expert attacker has just a slight impact, as shown in Table 6.

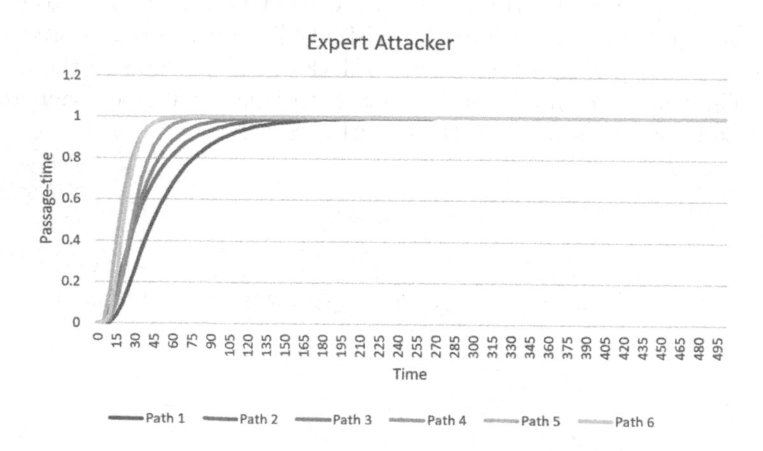

Fig. 13. Passage-time analysis of each path for the expert attacker when $p = 0.2$.

Table 5. The average time to compromise the system for beginner attacker.

Probability of exploit code availability	From *servU*5 to *completed* action	From *telnet* to *completed* action
$p = 0.2$	602 time units	465 time units
$p = 0.8$	237 time units	178 time units

Table 6. The average time to compromise the system for expert attacker.

The probability of exploit code availability	From *servU*5 to *completed* action	From *telnet* to *completed* action
$p = 0.2$	205 time units	156 time units
$p = 0.8$	180 time units	136 time units

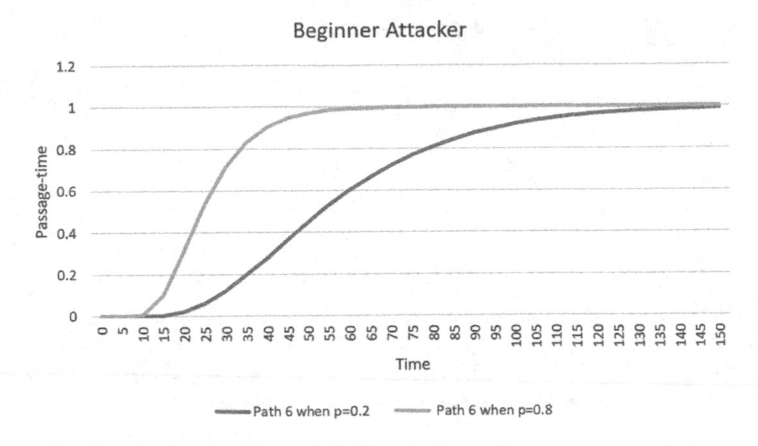

Fig. 14. Path 6 for beginner attacker when $p = 0.2$ and 0.8.

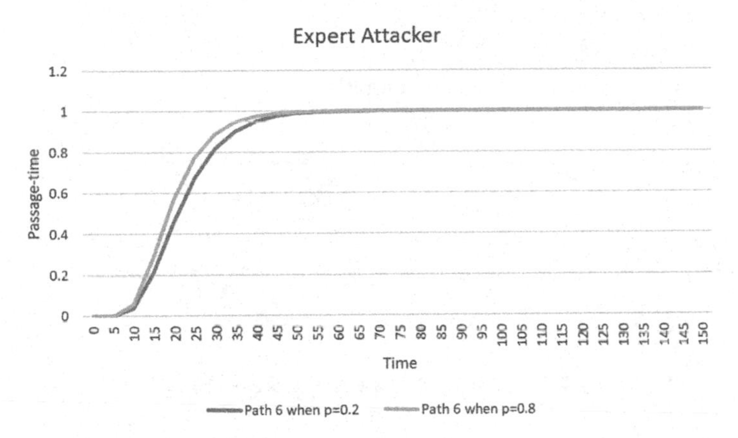

Fig. 15. Path 6 for expert attacker when $p = 0.2$ and 0.8.

Furthermore, Figs. 12 and 13 show the same risk order of the attack paths as in the first case study, as shown in Fig. 3. However, in this case study, we considered and implemented two important factors that impact the time to compromise the system. These factors are the probability of exploit code availability and attacker skill. Our evaluations of this case study clearly show the impact of these factors on the attacker time to compromise the system, as shown in

Figs. 12, 13, 14 and 15. In this case study, If an attacker fails to exploit the vulnerability, it returns to the first step in the attacker's steps graph to try again to attack the same host. Whereas in the first case study, when the attacker fails to exploit the vulnerability in any host, the attacker returns to the root node in the attack graph. This is clearly shown as an increase in the attacker time to compromise the system in the first case study, as illustrated in Fig. 3.

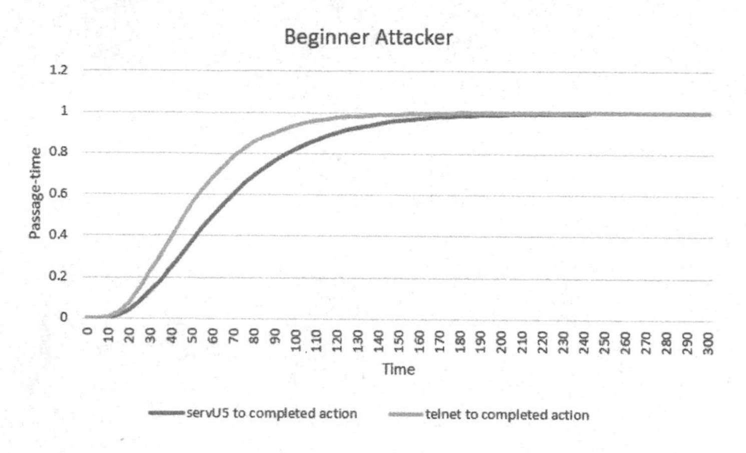

Fig. 16. Passage-time analysis for the beginner attacker when $p = 0.2$.

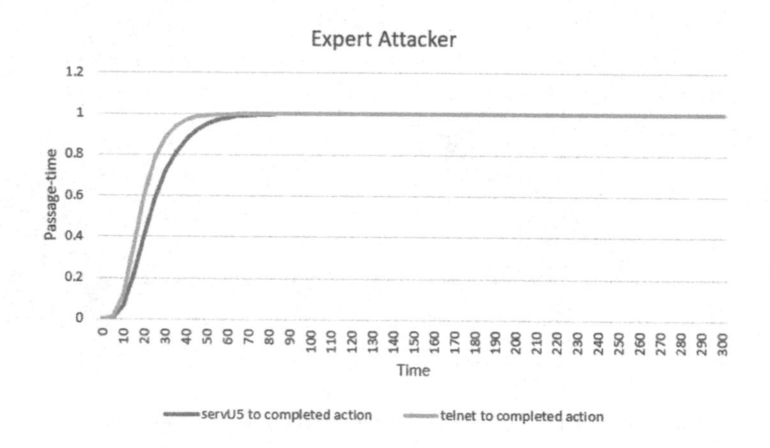

Fig. 17. Passage-time analysis for the expert attacker when $p = 0.2$.

6 Conclusion

In this paper, we proposed two methods to automate the generation of PEPA model based on a pre-existing attack graph specification with its existing vulnerabilities. The first method generates a PEPA model that comprises one sequential component representing the system and attacker coupled together. The second method generates a PEPA model that comprises two sequential components and a system equation. One component represents a system, and the other component represents an attacker.

We implemented the algorithms, and then we generated PEPA models for the case studies. We demonstrated through the case studies how we used the generated PEPA models to deduce the most and the least system security threatening paths and time to compromise, which the defender can use to indicate how much a safe time the system has before it is compromised. In addition, the model evaluation result can rank the risk of all attack paths.

In the first case study, we performed a sensitivity analysis by changing the probability of the vulnerability being breached of some actions and then did the passage-time analysis for each path. The model's evolution clearly demonstrates the sensitivity of each path to change and the effect on the attacker's time to compromise the system of each path. This can assist the defender in prioritising countermeasures.

In the second case study, we considered two types of attackers: beginner and expert. We proposed an attacker's steps graph for each type of attacker. Then, for each attacker, we generated the PEPA attack graph model. We estimated the attacker's time to compromise the system by taking into account the attacker's skill and the availability of exploit code. Then, we demonstrated how different attacker skills and the probability of exploit code availability influence the attacker's time to compromise the system. We also demonstrated how the lack of exploit code affects the time it takes a beginner attacker to compromise a system in comparison to an expert attacker. This study used the PEPA Eclipse plug-in to check the validity and to evaluate the PEPA models. The next step is to incorporate adaptive attacker and defender behaviours into the PEPA model, such as a learning behaviour. An attacker learning behaviour enables an attacker to learn from previous attack attempts, reducing the time required to compromise the system. Furthermore, as the defender learns about the attacker's behaviour and previous attack attempts, the defender will implement additional security measures to make it more difficult to compromise the system and to maximise the time it takes to compromise the system.

This paper has demonstrated the potential of using a stochastic process algebraic approach to model and analyse attack graphs. There is clearly a lot of scope for further work. The models presented are relatively small for the obvious reason of available space, however the methods used will scale to far larger attack graphs, unlike traditional CTMC based analysis. Furthermore, the analysis is quick to perform, which has the potential to facilitate a dynamic approach to analysing a live attack. Thus, an attacker's learning might be modelled in real time, allowing the defender to adapt the system protection accordingly. The rates used in the models in this paper are derived directly from the existing probabilities associated with attacker actions to exploit vulnerabilities. In future a more accurate depiction of time could be used to give better predictions of time to compromise.

References

1. Abraham, S., Nair, S.: A predictive framework for cyber security analytics using attack graphs. arXiv preprint arXiv:1502.01240 (2015)
2. Abraham, S.M.: Estimating mean time to compromise using non-homogenous continuous-time Markov models. In 2016 IEEE 40th Annual Computer Software and Applications Conference (COMPSAC), vol. 2, pp. 467–472 . IEEE (2016)
3. Garg, U., Sikka, G., Aawsthi, L.K.: A systematic review of attack graph generation and analysis techniques. In: Computer and Cyber Security: Principles, Algorithm, Applications, and Perspectives, pp. 115–146. CRC Press (2018)
4. Hillston, J.: A compositional approach to performance modelling, no. 12. Cambridge University Press (2005)
5. Hillston, J., Ribaudo, M.: Stochastic process algebras: a new approach to performance modeling. Modeling and Simulation of Advanced Computer Systems. Gordon Breach (1998)
6. Kaluarachchi, P.K., Tsokos, C.P., Rajasooriya, S.M.: Cybersecurity: a statistical predictive model for the expected path length. J. Inf. Secur. **7**(3), 112–128 (2016)
7. Khaitan, S., Raheja, S.: Finding optimal attack path using attack graphs: a survey. Int. J. Soft Comput. Eng. **1**(3), 2231–2307 (2011)
8. Lallie, H.S., Debattista, K., Bal, J.: A review of attack graph and attack tree visual syntax in cyber security. Comput. Sci. Rev. **35**(2020), 100219 (2020)
9. Leversage, D.J., Byres, E.J.: Estimating a system's mean time-to-compromise. IEEE Secur. Priv. **6**(1), 52–60 (2008)
10. McQueen, M.A., Boyer, W.F., Flynn, M.A., Beitel, G.A.: Time-to-compromise model for cyber risk reduction estimation. In: Gollmann, D., Massacci, F., Yautsiukhin, A. (eds.) Quality of Protection. Advances in Information Security, vol. 23. Springer, Boston, MA (2006). https://doi.org/10.1007/978-0-387-36584-8_5
11. Pokhrel, N.R., Tsokos, C.P.: Cybersecurity: a stochastic predictive model to determine overall network security risk using Markovian process. J. Inf. Secur. **8**(2), 91–105 (2017)
12. Sun, F., Pi, J., Lv, J., Cao, T.: Network security risk assessment system based on attack graph and Markov chain. J. Phys. Confer. Ser. **910**, 012005 (2017). IOP Publishing (2017)
13. Swiler, L.P., Phillips, C., Gaylor, T.: A graph-based network-vulnerability analysis system. Technical Report. Sandia National Labs, Albuquerque, NM (United States) (1998)
14. Zheng, Y., Lv, K., Hu, C.: A quantitative method for evaluating network security based on attack graph. In: Yan, Z., Molva, R., Mazurczyk, W., Kantola, R. (eds.) NSS 2017. LNCS, vol. 10394, pp. 349–358. Springer, Cham (2017). https://doi.org/10.1007/978-3-319-64701-2_25

Towards Calculating the Resilience of an Urban Transport Network Under Attack

David Sanchez(✉) and Charles Morisset

Newcastle University, Newcastle upon Tyne, UK
D.E.Sanchez-Oliva2@newcastle.ac.uk, charles.morisset@ncl.ac.uk
http://www.ncl.ac.uk

Abstract. In this article we present a methodology to calculate the resilience of a simulated *cyber-physical Urban Transport Network* (UTN) under attack. The UTN simulation was done using CompRes, a software that we created that uses Monte Carlo method on a uni-dimensional cellular automaton. We quantified resilience in a novel way by considering it not as "avoiding deviation from the target performance function", but as the "avoidance of the performance collapsing limits of the system whilst under attack". We show the use of our proposed resilience approach by analysing two simulated UTN attacks scenarios (link attacks and decision attacks) using CompRes. We believe our findings are general enough to be applicable to similar dynamical systems and in some circumstances easier to apply than traditional measurement. We discuss its advantages and disadvantages, usefulness in security, monitoring of dynamical systems and artificial intelligence.

Keywords: Resilience · Dependability · Cyber-Physical systems · Computer Security · Urban Transport Network · Dynamical Systems · Cellular Automata · Stochastic Modelling · Monte Carlo Simulations

1 Introduction

Resilience is easier to use than to define [16,20,23]. The term is usually treated as an *operationalisable* concept because even though there is a lack of consensus on what exactly means, this is not perceived as an obstacle to use it to produce results [20]. Despite this, it is always preferable to have a clear definition, even if it is not the most accurate, and metrics are key to diminish ambivalence [5].

In this article we present a methodology to quantify the resilience of a simulated Cyber-Physical Urban Transport Network (UTN) under attack. This use case gave us interesting ideas about how to define and quantify resilience that could be applicable to other dynamical systems. In general terms, rather than interpreting resilience as "avoiding deviation from expected performance", we consider it as "avoidance of collapsing limits whilst under attack".

We also present a software, called CompRes, that we created to test our hypothesis. It uses a uni-dimensional cellular automaton to synthetically recreate the typical stochastic behaviour found in UTNs. This simulator runs iterative

M. Forshaw et al. (Eds.): PASM 2022, CCIS 1786, pp. 27–46, 2023.
https://doi.org/10.1007/978-3-031-44053-3_2

scenarios (Monte Carlo method) to create a visual region of expected output. The resulting region is used to interpret a resilient system as one that manages to stay within these safe boundaries of performance whilst under attack.

Our approach is novel because of four reasons. First, in computing resilience is typically reduced to *availability*. This simplification might be enough to many use cases, but it seems insufficient when considering cyber-physical units (similar opinion at Bishop et al. [5]). These systems deal with a higher level of tangible dynamics than traditional computers and often find themselves in scenarios in which they are "available" but with serious diminished possibilities. Second, typical considerations of resilience as deviation from optimal performance do not consider the difficulty of calculating the optimal performance. Third, most resilient definitions rightfully consider keeping minimal quality-of-service as a resilience indicator but fail to address output excess as a resilience issue. And fourth, other resilience definitions do not combine our reasons 1, 2 and 3 into a cohesive measurable definition.

The next sections of this article present our approach organised as follow. Section 2 covers minimal background to understand the context. Section 3 presents our formal definition of resilience. Section 4 introduces our model of performance (produced by CompRes). Section 5 shows our quantification of resilience. Section 6 shows under-attack scenarios and corresponding analysis (using CompRes). Section 7, 8 and 9 provide discussion, further paths, and conclusions respectively.

2 Background and Related Works

Urban Transport Networks (UTNs) are our dynamical system use case. UTNs refer to the infrastructure that allows the delivery of commodities across cities [2,3]. They are a subset of Critical Infrastructure Systems (CISs) [10,18] that delivers vehicles (rails, roads, distributors, streets, airports, docks, etc.), water (primarily in form of aqueducts), power (by electric grids, gas pipelines, petrol distribution, gasoline supply, etc.) and information (telecommunication infrastructure, fibre optic, cellular antennas, satellites, etc.).

UTNs resilience is important because of the consequences when they malfunction. On best-case scenarios, non-resilient UTNs produce deficient but tolerable services (e.g. [6]); on worst-case scenarios, they cause human loses (e.g. [24,30]). It is expected that UTNs resilience may decline in coming years [21,22].

UTNs are interesting to computing science because they are becoming cyber-physical systems and the computerisation of those systems without carefully considering resilience can create powerful but brittle systems, e.g., 2003 Italy blackout [6].

Resilience Definition and Measurement can be done directly, e.g., availability of a system, or indirectly by the effect that causes on a system (Some Biological samples in [16]). Adversarial events that trigger the resilient behaviour of a system are either modelled as random errors or as coordinated attacks. Random errors are typically used to study inner imperfections of a system and are expressed with probabilistic measurements. Coordinated attacks, on the other hand, model external influences and have context-dependant measurements. We are studying coordinated attacks as our adversarial effect on dynamical systems (UTNs). An example of a natural coordinated attack to a UTN is rain.

A generally well accepted non-ambivalent definition of resilience for dynamical systems is provided by Ouyang and Dueñas-Osorio [19]:

$$R(T) = \frac{\int_0^T P(t)dt}{\int_0^T TP(t)dt} \tag{1}$$

Equation 1 presents resilience as the quotient of two functions that accumulate performance over time. The target performance $TP(t)$ represents the level of performance that should be acquired to guarantee a 100% of *quality of service* and $P(t)$ is the actual performance either under the influence of an adversity or not. Maximum resilience is only possible when $P(t) = TP(t)$. No resilience means $P(t) = 0$.

3 Performance and Resilience on Dynamical Systems

The *performance* of a dynamical system is usually given by a specific metric that relates the system's input to its output. For instance, for a road network, the performance might be measured as the average velocity over the network over time, or as the maximum number of vehicles reaching the destination in a certain time depending on the density of the network (traffic flow).

Given a system, we need to distinguish between its *actual performance*, which is normally an observable measure for a concrete or simulated system, and its *expected performance*, which is a specification or set of requirements that express how the system should behave under normal conditions.

We consider here that a system will have a normal amount of *variation*, i.e., given similar inputs, the system might produce slightly different outputs. For instance, on a road network, the average velocity will depend on the number of vehicles (which is the input) but also on the fact that, in practice, not every driver chooses to drive at the same velocity, even under identical conditions. We therefore consider that the expected performance of a system is given as a *minimum expected performance* and *maximum expected performance*, and we say that a system is performing as expected if and only if its actual performance is within the minimum and maximum expected performance under normal variation. However, in some cases, abnormal amounts of variation can occur, which we denote as *attacks*. For instance, in the road network, an attack could be the situation where a high number of cars suddenly stop for a few minutes and then accelerate at full speed (which could be the result of an attacker controlling self-driving vehicles). We say that a system is *resilient against an attack* if and only the performance of the system under attack is within the minimum and maximum expected performance of the system without the attack. In other words, a system is resilient against an attack if this attack does not negatively impact its performance beyond the expected boundaries, although it is possible that the actual performance of the system under attack will be lower than the performance of the system without the attack.

Finally, we consider that a given *modification* of a system is an effective *defence against an attack* if, and only if, the modified system is resilient against this attack.

4 CompRes: Modelling UTN Performance

In this section we present a software, called *CompRes*, that we created to study the resilience of a simulated UTN under attack. CompRes is a UTN Monte Carlo Simulator that in the current version models the real-life complex dynamics that would be produced by the flow of autonomous vehicles (robot-cars) that travel from point A to point B under the influence of some attacks.

CompRes is written in JavaScript and runs on web browsers. We are releasing the source code as open source[1]. The interaction is done through a form that collects parameters (Table 1) and outputs simulations (Fig. 1) and statistics (Details at the end of the section).

CompRes is envisioned as a tool to test resilience/attack scenarios either normal/attack or attack/attack. As it can be seen in Table 1, it collects variables used in both scenarios, as well as those only used by control or attack. It models UTNs assuming that these could be reduced into *position/decision* relationships. In this version, *position* is simplified as an array and *decision* is provided by a plug-in that uses the NaSCh vehicular traffic simulation rules [17]. This version of CompRes therefore produces similar results to a typical NaSCh simulator in a closed-circuit loop; however, the arrangement of concerns (position/decision) allows future versions to use different deciders besides the NaSCh one. CompRes does not focus on traffic flow realism[2] but on comparing hypothetical attack scenarios. This makes CompRes the early prototype of an extensible and versatile tool that could even mimic commodities flow that do not follow vehicular traffic rules (e.g., water pipelines, power grids, parcels delivery, etc.).

4.1 Simulation

Each simulation produced by this version of CompRes (Fig. 1) creates the traffic that will be experienced by a series of robot-cars that circulate on a road. As in real life, each vehicle will try to travel at maximum allowed velocity (distance/time). Each simulation represents a working UTN scenario (normal or under attack) and is modelled as a uni-dimensional cellular automaton $U = (\Sigma, N, \lambda, M, S)$. This section describes each component.

[1] Code available at https://github.com/david-San/CompRes. CompRes can be seen in action at https://david-san.github.io/CompRes/public_html/index.html.

[2] There are better tools to study traffic flow such as MovSim https://traffic-simulation.de/ring.html [28].

Table 1. CompRes Input. Parameters Details.

Parameter	Explanation	Scenario	Values	e.g
Number of cells	Length of the road. Cell size is an average car (3.5 m)	Both	10, 600	60
Frames Number	Number of frames on each simulation. Each frame is a photo of the road	Both	10, 20	10
Initial Frames to Discard	Frames out of statistics. Simulation typically stabilises after 10 initial frames	Both	10, 20...	10
Density Init	Starting point for density steps	Both	[0.0 - 1]	0.01
Density End	Ending point for density steps	Both	[0.0 - 1]	0.5
Density Steps	Total amount of densities	Both	1, 2, 3	20
Plot Maximum Y value	Plot Y-axis maximum value	Both	100	100
Plot Minimum Y value	Plot Y-axis minimum value	Both	−0.5	−0.5
Movable Max Speed	Velocity limit for a robot-car	Both	[1 - 5]	5
Movable Performance High Limit	Robot-car theoretical maximum velocity	Both	[1 - 5]	5
Movable Performance Low Limit	Robot-car theoretical minimum velocity	Both	[1 - 5]	0.5
Number of simulations	Monte Carlo method. Higher creates more precise resilience regions	1 (Control)	10, 20	10
Probability Pushing Brakes Randomly	Higher braking means more unforeseen issues on the road (Attacks)	1 (Control)	[0 - 1]	0.1
Hacked Movables	Number of cars with hacked decisions	1 (Control)	0, 1, 2	0
Number of simulations	Monte Carlo method. Higher creates more precise resilience regions	2 (Hacked)	10, 20	10
Probability Pushing Brakes Randomly	Higher braking means more unforeseen issues on the road (Attacks)	2 (Hacked)	[0 - 1]	0.5
Hacked Movables	Robot-car theoretical minimum	2 (Hacked)	0, 1, 2	1

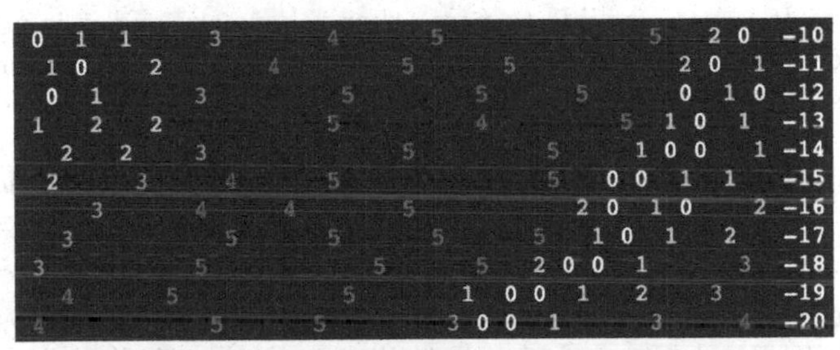

(a) Velocity of each robot-car frame by frame. Velocity (0 min, 5 max) highlighted to show congestion.

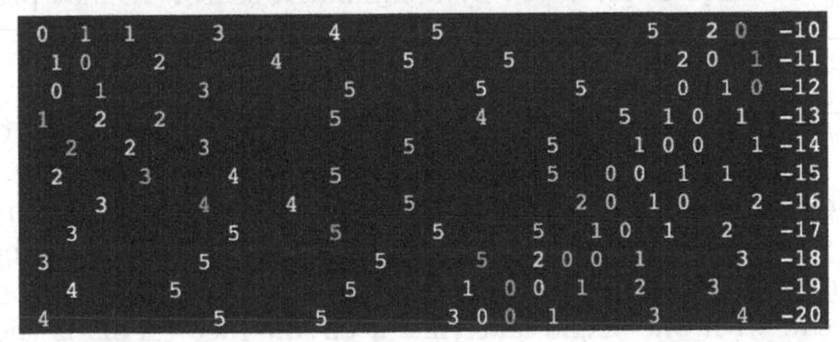

(b) Velocity of each robot-car frame by frame. Robot-cars highlighted to track them through frames.

Fig. 1. CompRes Output. Traces Sample at 0.1733 density. Each line is a σ_i frame (Showing frames 10...20). (Dataset 1 Sim. Control 3; $d = 0.1733$. [26]).

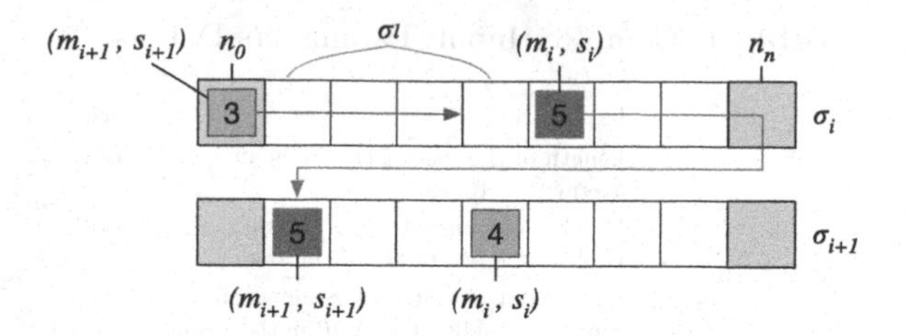

Fig. 2. Cellular Automaton. When a movable reaches the end of the array is placed at the corresponding initial position. (m_i, s_i) represents m_i movable and s_i associated stakeholder. Numbers represent velocity. σ_l is the distance sensor. n_i are the linked regions (n, λ).

Trace (Σ). The whole simulation is a map of discrete steps $(\sigma_i \in \Sigma)$. Each map entry stores $(i, (N, \lambda, M, S)_{\sigma_i})$. Each σ_i represents a photograph (frame) of the road at a given discrete time (In Fig. 1, frames are marked with -0, -1, -2...). Once the `Run` button is pressed, σ_0 is generated randomly with a normal distribution of robot-cars (M) on a road (N, λ) according to the densities parameter. Densities are specified with a starting and ending point as well as steps. Subsequent σ_i are produced by a loop that traverse the road and activates each robot-car driver (stakeholder) to take a decision about the new position of the car. Each stakeholder uses the distance reported by a simulated sensor (σ^l) to take the decision (Fig. 2). It is only possible for a car (m) to occupy only a single cell (n) at each σ_i. It is not possible to pass other robot-cars and it is not possible to crash. The whole environment is produced by the parameter *Frames Number* that specifies how many σ_i will be generated. The parameter *Initial Frame to Discard* is used to delete the initial frames in which robot-cars are starting to move. CompRes allows tracing for a car across σ_i since each one is identifiable by a colour (See Fig. 1).

Movable Commodities (M). A car $(m \in M)$ is a JSON object that store identification, position, velocity and visual attributes. Each m is linked to its driver stakeholder S by the method `stakeholderDecide(nextMovableDistance)`. `nextMovableDistance` is produced by a simulated sensor σ^l that provides the distance $d_m = n_{m+1} - n_m$ (Fig. 2).

Region (N, λ). A road, represented by an array of N cells. Each cell of the array is $n \in N$. The continuity of those n that are necessary to travel from one to another (expressed by λ) is granted by the continuous array address space. For simplicity, we used a circular structure, so accessing the position beyond the last array index means accessing the first array position (Fig. 2). This mimics a closed ring road with one single lane (like a circuit race). This is the simplest model typically explored in vehicular simulation [28].

Stakeholder (S). Robot drivers. Each $s \in S$ has a method `decide` to determine the car behaviour.

4.2 Stakeholders Decision

Stakeholders' agreement to move or not is provided by a plugin that uses rules proposed by Nagel and Schreckenberg [17] in the NaSch micro-simulator for traffic. Despite the rules simplicity, NaSch simulators have been demonstrated to produce realist results for vehicular traffic [1, 17]. We used the revisited rules presented in Schadschneider [27].

The process of agreeing to move m was complemented by the action of accelerating or decelerating m. Therefore, the position n_m is a consequence of the velocity $v_m \in 0, 1, 2, ..., v_{max}$. The gap between the mth movable and the next is d_m returned by a simulated sensor (σ^l). Each stakeholder s_m will decide how to drive its movable according to these NaSch rules [27][3]:

1. Accelerate: If $v_m < v_{max}$, $v_m \to min(v_m + 1, v_{max})$.
2. Decelerate: If $d_n \leq v_m, v_m \to min(v_m, d_m - 1)$.

Step 1 models the preference to drive fast; step 2 prevents collisions; Once s_m takes the decision, m recalculates its position using $n_m \to n_m + v_m$.

Our implementation covers $s : M \times N \times N \times \Sigma \to \{\perp, 0, 1\}$ since, it is possible that given v and d_m, s_m could decide to accelerate or decelerate using rules 1 and 2 ($\{1\}$) or not to do anything when rule 2 annuls rule 1 ($\{0\}$). For simplicity, we left out the case when s_m is unable to decide ($\{\perp\}$).

The decision process is isolated and therefore easily replaceable. Since we are using robot-cars, typical considerations studied by other authors [28] such as politeness, bias, stress, speed of human reactions by age, etc. are irrelevant.

4.3 Simulation Results: Performance Metrics

Each scenario analysed using CompRes returns robot-cars velocity over time in different formats. It is possible to see summary results of the flow of all robot-cars together as well as detailed information on each simulation and even on each robot-car.

CompRes summarises the results of the simulations at different densities in charts that present (movables speed)/density (e.g., Fig. 3a) and interpolated graphs that show tendencies of (movables flow)/density behaviour (e.g., Fig. 3b). Such Speed-Density-Flow relationships are difficult to calculate empirically [28].

Detail simulation information is similar to Fig. 1 and also includes in-detail graphs for each robot-car velocity performance (e.g., Fig. 4a and b) with some statistical information similar to Listing 1.1 that will be further explained in Sect. 6.3. CompRes allows dynamic high-lighting of each robot-car. All figures (Figs. 1, 3 and 4) and Listing 1.1 are part of the same CompRes run, they have been divided for explanation purposes.

[3] Note that the actual definition of NaSch rules includes another element, which will be introduced in Sect. 6.

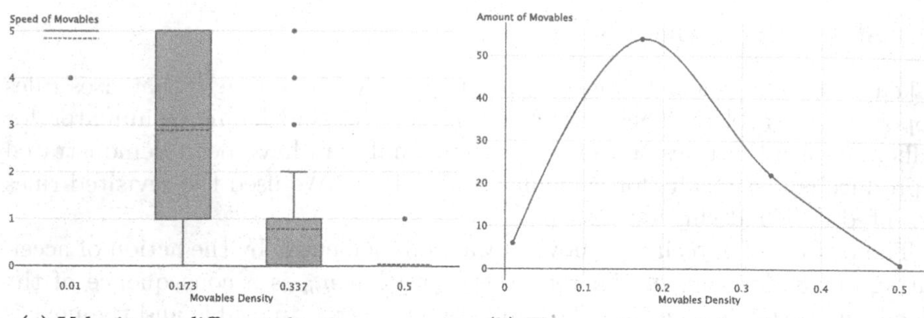

(a) Velocity at different density. (b) Flow as the road density increases.

Fig. 3. CompRes summary output sample (Dataset 1 Sim. Control 3 [26]).

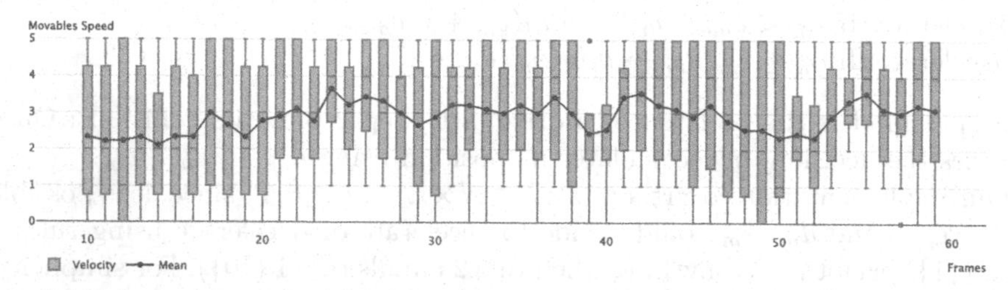

(a) Robot-cars velocity (60 frames; 50 cells; $p = 0.1$; density=0.1733)

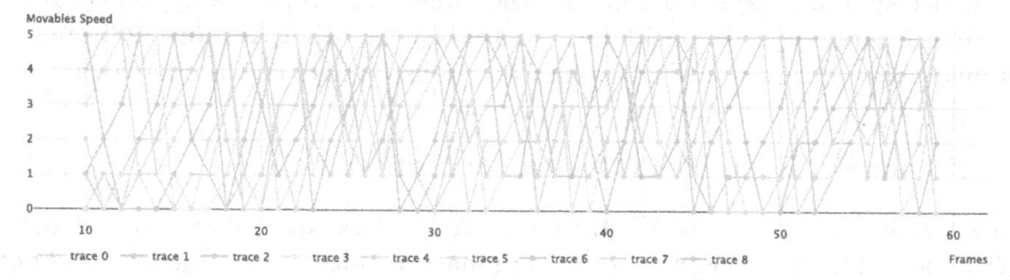

(b) Each Trace is one robot-car (60 frames; 50 cells; $p = 0.1$; $d = 0.1733$).

Fig. 4. CompRes details output (Dataset 1 Sim. control 3; $d = 0.1733$ [26]).

```
Number of Cells: 50
Density: 0.17333333333333334
Number of Movables: 9
Number of Frames: 60
Total Movables Crossed Finish Line: 54
Total Movables Crossed Finish Line per Frame: 0.9
Initial Frames to Discard: 10

Speed average: 2.4092592592592594
Speed Max Boundary for Movable: 5
Speed Min Boundary for Movable: 0.5
Resilience Speed Cumulative Index relative to Max Speed (Should be <1): 0.5782222222222223
Resilience Speed Cumulative Index relative to Min Speed (Should be >1): 5.782222222222223

Flow for this density: 54
Flow Max Boundary Region for this density: 76
Flow Min Boundary Region for this density: 54
Resilience Flow Cumulative Index relative to Max Flow (Should be <1): 0.7105263157894737
Resilience Flow Cumulative Index relative to Min Flow (Should be >1): 1
```

Listing 1.1. Fig. 4 statistics (Dataset 1 Sim. Control 3; $d = 0.1733$. [26])

The performance metrics of each movable(m) is $P_m(t) = \frac{\text{Robot-car Speed}}{\text{time}}$. Movables velocity has a scale from 0-5 (low to maximum velocity respectively) to be compatible with NaSch simulations (See Sect. 4.3).

The performance metrics of each simulation is traffic flow $P_\sigma(t)$. *Traffic flow*[4] is defined as the number of vehicles (M) passing a predefined location (x) per time unit (t) [28]: $P_\sigma(t) = \frac{M_x}{t}$.

CompRes establishes the predefined location (x) at the end of the simulated road (Before the loop that takes movables to the beginning). The discrete time unit of the simulator is the frame; however, since all simulations produced by an experiment have the same numbers of frames, we simplified n frames $= 1$ simulation. Consequently, the performance $P_\sigma(t) = M_x$

Validation. CompRes NaSCh plug-in produces similar laminar vehicular flow as a typical NaSCh traffic simulator [17]. This can be appreciated in the similarities of Figs. 3b and 5b, and Figs. 1 and 5a. Since NaSch simulations have been validated using aerial photography [17] and street sensors [1], we assume by transitivity that CompRes produces realistic vehicular flow, and it is useful to study realistic scenarios.

5 Quantifying Resilience on Dynamical Systems

Measuring resilience using Eq. 1 presents some limitations. The most evident is that at a given t, TP might not be a value but a range since dynamical systems are built considering levels of tolerance that change over time (examples of these decay functions by age in [7,15,31]).

Let $P(t)$ be the actual performance of a system at a given time. This performance might be under the influence of adversity, called *attacks*, that may alter

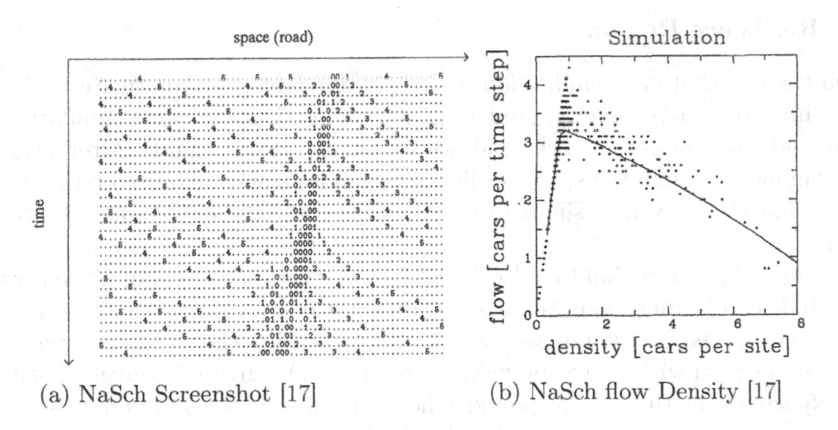

(a) NaSch Screenshot [17]

(b) NaSch flow Density [17]

Fig. 5. Original NaSch simulator results [17]. As it can be seen they are like what we obtained in Fig. 3 and Fig. 1 respectively.

[4] Equivalent to throughput in other contexts such as computing.

its performance. These attacks influence the system to perform either above or below its intended outcome if the system had no attacks. $LP_{min}(t)$ is the lowest possible performance for such system to be usable and $LP_{max}(t)$ is the maximum possible performance for that system to operate safely.

We define the cumulative minimum and maximum resilience index, R_{min} and R_{max}, up to a time T as follows:

$$R_{min}(T) = \frac{\int_0^T P(t)dt}{\int_0^T LP_{min}(t)dt} \quad (2) \qquad R_{max}(T) = \frac{\int_0^T P(t)dt}{\int_0^T LP_{max}(t)dt} \quad (3)$$

And the notion of resilience as follows:

$$R(T) = \begin{cases} \text{Resilient} & \text{if } R_{min}(T) \geq 1 \text{ and } R_{max}(T) \leq 1 \\ \text{Non-resilient} & \text{otherwise} \end{cases} \quad (4)$$

Therefore, a system is *resilient* when it does not cross its performance limits whilst operating under the influence of attacks during a time frame.

Equation 4 may look at first sight as Eq. 1 but has different semantics. It substitutes TP by a range (LP_{max} - LP_{min}) and provides a binary characterisation of resilience and non-resilience. It also allows designers to provide LP_{max} and LP_{min} (sometimes easier than finding TP) that could have complex variation of performance tolerance over time. Figure 6 shows a graphical explanation.

6 Scenario Analysis

6.1 Resilience Region

Given the fact that each simulation is slightly different from another one, CompRes simulates finding LP_{max} and LP_{min} by repeating the same simulation of traffic under normal non-attack conditions (Monte Carlo method). After a significant number of simulations, the resilience region is clearly delimited (stabilises). This means that any new simulation will stay within the determined resilience region.

Figure 7 shows a region found with 100 simulations of a UTN of 60 cells road with 20 densities. Increasing the number of iterations produces a more accurate ample boundaries (Maximum and minimum). Increasing the number of densities produces a region with higher boundary resolution. At any of the given densities, it is expected that the system provides flow within the found boundaries.

Figure 8a is a practical example of the theoretical explanation presented in Fig. 6. The robot-cars acceleration/deceleration behaviours make the traffic flow to stay within the collapsing boundaries in a very unpredictable way. In fact, in

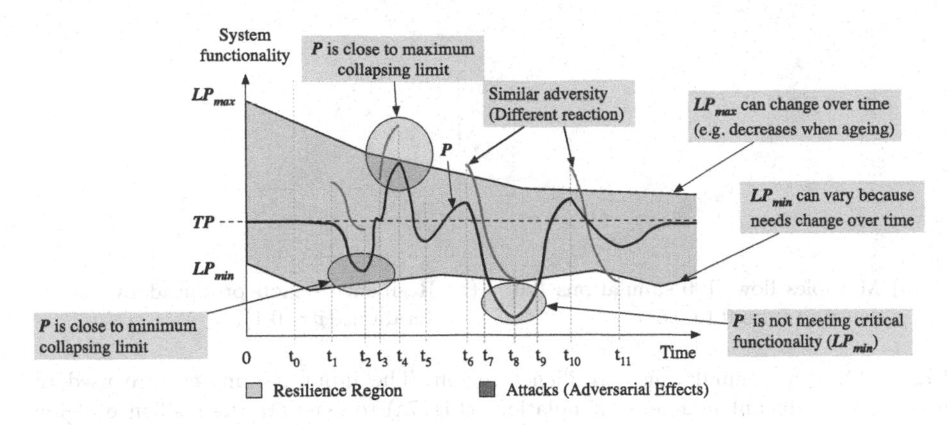

Fig. 6. Resilience Sample. First attack is absorbed with no consequences (t_0). Second attack reduces performance but system is still acceptable (t_1, t_2). System compensate reduction in performance with extra effort (t_2, t_3). Third attack happens but the system absorbs it even though is closed to collapse (t_3, t_4). System recovers (t_4, t_6). Attack makes the system drop below $LP_{min}(t_6, t_8)$. System is not resilient (t_7, t_9). System recovers (t_9) and for the next attack (t_{10}, t_{11}) that is similar to the previous one, it adapts and does not collapse.

Fig. 8b we can see that two different simulations may have completely different behaviours and still be valid under the normal expected performance of the system.

Each simulation creates a slightly different result thanks to:

- A randomise normal distribution of the vehicles done at σ_0 when starting each simulation.
- A configurable stochastic parameter p that produces random brakes. p will be covered in Sect. 6.2.

6.2 Attack Scenarios

We can classify the possible attacks to our model as follows:

- Physical Attacks. Attacks on the integrity of the road (N, λ), integrity of movables M and integrity of sensors σ. We are not covering these kinds of attacks since they are usually approached when resilience is considered *availability* and have been studied by other authors [4,11,13,25,29]. e.g., a Tsunami (extreme flood) that destroys roads and vehicles is a typical attack on this kind.
- Link Attacks. Attacks that affect the physical flow of movables without destroying physical integrity. e.g., heavy rain that floods a road, but it does not destroy it. Eventually water is dispersed, and the road is usable again without repairs.
- Decision Attacks. Logical attacks on the decision process that is carried out by S. However, a bad decision may collapse links or destroy physical infrastructure.

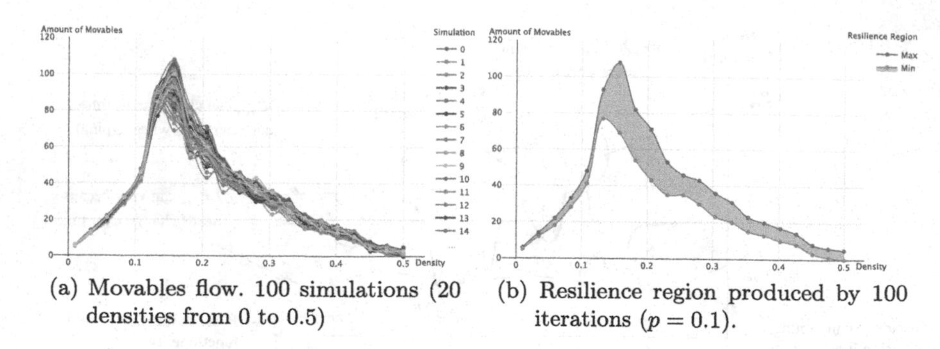

(a) Movables flow. 100 simulations (20 densities from 0 to 0.5)

(b) Resilience region produced by 100 iterations ($p = 0.1$).

Fig. 7. CompRes simulating a resilience region. The input parameters are used to produce a significant number of simulations (Fig. 7a) to generate the resilience region of the system (Fig. 7b). (Dataset 2 [26]).

Besides the obvious self-attack that is produced naturally by the system when density varies, CompRes simulates link and decision attacks.

Link Attacks. These are simulated by adjusting the p parameter. p was originally defined in the NaSch decision process in a third rule to introduce asymmetry between acceleration and deceleration:

3. Randomise: If $v_m > 0, v_m \rightarrow max(v_m - 1, 0)$ with probability p.

This asymmetry involves for example, typical differences in the performance of the engine vehicles and brake mechanisms but also any kind of small imperfection that can be found in the road. Consequently, p could be used to provide road adversity (attacks) to the system, i.e., a higher use of the brakes of a vehicle

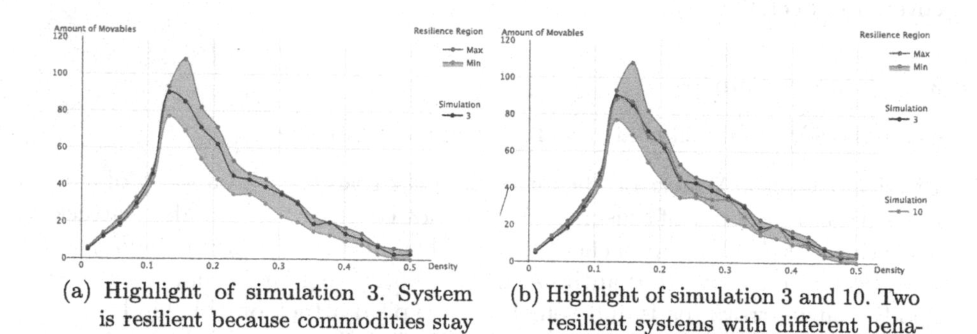

(a) Highlight of simulation 3. System is resilient because commodities stay within the accepted performance region (Between Min and Max flow).

(b) Highlight of simulation 3 and 10. Two resilient systems with different behaviours operating within accepted performance boundaries.

Fig. 8. CompRes simulation results sample. Figure 8a highlights how a typical movable (Simulation 3) stays within the resilience boundaries. CompRes allows highlighting each simulation for verification. Figure 8b shows movables with different behaviour operating within resilience boundaries. (Dataset 2 [26]).

because there is an increase adversity on the road would simply mean a higher p value. Also, since p is a randomised parameter, it is a good fit for simulating unplanned adversarial effects.

Figure 9 shows a typical output of this attack on CompRes. Figure 10b shows the system under 100 attack simulations. Figure 10c shows a comparison between the resilient region and a region draw by the attacks. Simulation 23 shows an instance of the system under attack that manages to recover above 0.45 density; however, at that point the system may already be found intolerable by users.

Decision Attacks. These are simulated by substitution of the logic of the stakeholder. This kind of attacks are particularly important to us since we are simulating cyber-physical movable commodities (robot-cars) and interfering the decision process is popular by hackers (as evidenced by [9]).

To produce the attack, we created a `stakeholderHack1` with a hacked acceleration rule: Once the car reaches speed 3, the car will go to 0 (simulating a turn off) with a probability 0.5 regardless of the distance in front of the car:

1. Accelerate: If $v_m \leq 3$, $v_m \rightarrow min(v_m + 1, v_{max})$, else $v_m \rightarrow 0$ with $p = 0.5$.

The other rules are kept with no modifications. This behaviour aims to simulate a hacked transmission vehicle, such as the one presented in Greenburg [8] that was remotely compromised while being driven at 70 mph. We are not pursuing a full stopped vehicle since it is clear the consequence on this simulator (full stall), instead we are assuming that the hacker does not want to be easily discovered while providing maximum annoyance. We think that our hacked stakeholder would present an anomaly that would be difficult to detect in common driving conditions and even harder to diagnose by a mechanic.

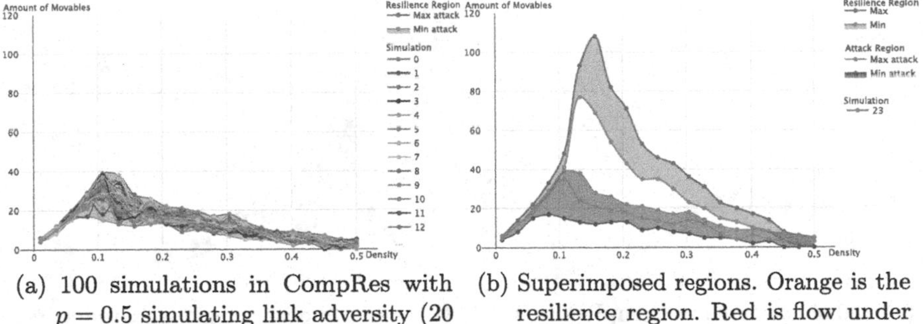

(a) 100 simulations in CompRes with $p = 0.5$ simulating link adversity (20 densities from 0 to 0.5).

(b) Superimposed regions. Orange is the resilience region. Red is flow under attack region. Simulation 23 shows an instance flow under attack.

Fig. 9. A sample of CompRes simulation results showing a simulated attack on link layer. The attack incapacitates the system until Approx. 0.4 density. Figure 9a shows 100 stochastic consequences. $p = 0.5$ means lots of road adversity. Figure 9b shows the attacks are out of the resilience region. System is resilient before 0.1 and after 0.4 but barely usable. (Dataset 2 [26]).

As it can be seen in Fig. 10, this attack is not as evident as the one presented in Fig. 9. For some densities and some simulations, the attack is effective since the attack region is outside the resilience zone, but because a significant part of the attack region overlaps the resilience region, the system was resilient in many of the attacks. The highlighted simulation 11 shows a case in which the system recovers from an attack after density 0.2. The attack happens between 0.1 and 0.2 approx.

6.3 Quantifying Resilience Under Attack

As it can be seen in Figs. 9 and 10, using our definition is straightforward to see when the movables flow is out of the expected resilience region and consequently when the system is resilient and when not.

It is important to highlight that Figs. 9 and 10 show a static vision of the density change but in real life road density is continuously changing. Therefore, the upper and lower boundaries of the resilience region would change dynamically. A movable that enters such a dynamical system needs to be compared and measured against the lower and upper boundaries that are relative to the density of the time it travels through the system.

CompRes calculates the resilience index of each movable, considering the resilience region that has been calculated. Each simulation is accessed by a button that reveals all information about the movable.

Once the two resilience indexes (Eqs. 2 and 3) have been calculated for each movable commodity (robot-car), it is possible to determine, for example which of these two movables is more resilient than the other. It is important to note that when the resilience index is calculated, it loses detail, and we could end up with two systems that managed to have the same resilience index, even though, one of the movables, had a much bigger stress than the other.

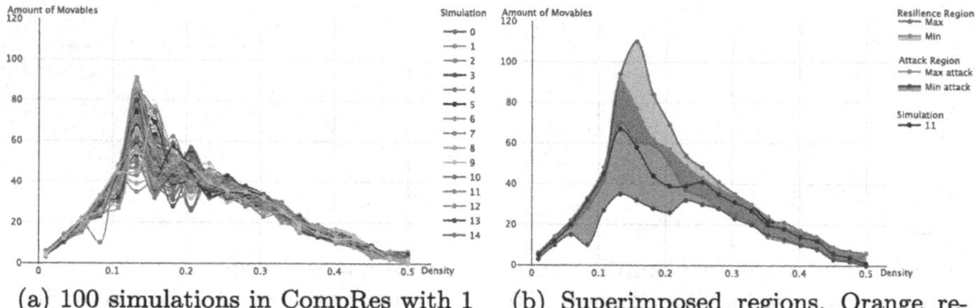

(a) 100 simulations in CompRes with 1 hacked stakeholder (20 densities from 0 to 0.5; $p = 0.1$).

(b) Superimposed regions. Orange region (almost covered) is resilience region (Same Fig. 10b). Red is under attack region. Simulation 11 shows flow under attack.

Fig. 10. CompRes simulation showing an attack on stakeholder. The attack is effective between 0.1 and 0.3 but not always. One way to interpret this phenomenon is that the system is resilient to the attack. (Dataset 3 [26]).

As an example, let us consider Fig. 10c simulation 11 at density 0.1815 there is clear evidence of an attack on the system (out of the resilience region). The system has a flow of 45 movables which is under the expected minimum resilience of 52 (and obviously maximum 83). The resilience indexes (R_{min} and R_{max}) can be calculated using the Eqs. 2 and 3:

$$R_{min}(T) = \frac{\int_0^T P(t)dt}{\int_0^T LP_{min}(t)dt} = \frac{44}{59} = 0.74 \quad R_{max}(T) = \frac{\int_0^T P(t)dt}{\int_0^T LP_{max}(t)dt} = \frac{44}{84} = 0.52$$

R_{min} should be ≥ 1 and it is 0.74. R_{max} should be ≤ 1 and it is 0.52. These results show that the system under attack has not saturated the maximum boundary region for the system, but it is clearly not reaching the minimum value to be effective.

7 Discussion

The analysis of the stochastic behaviour of a dynamical system, such as a UTN, offers many possibilities to safeguard the system. As it can be seen, these systems have a behaviour that is dynamical but to some degree predictable. We have shown that we can use a simulator not intended for studying resilience (NaSCh) as a viable tool to do so by creating the *resilience region*. This methodology can be used to detect attacks on those systems.

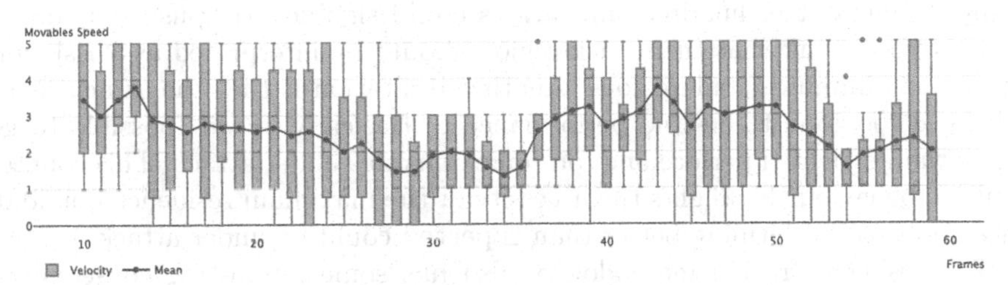

(a) Average speed of all vehicles when 1 hacked vehicle is present.

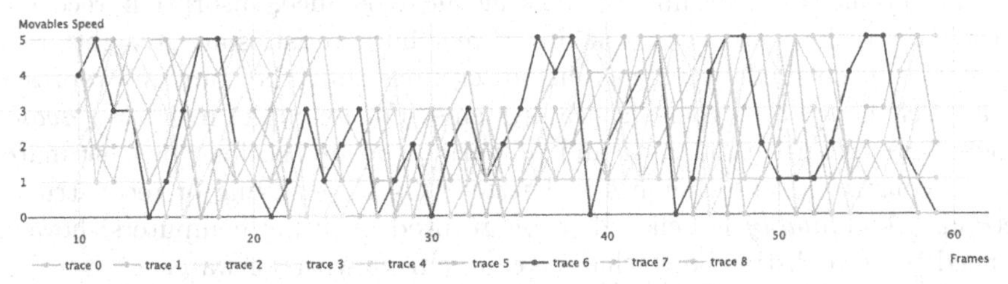

(b) Speed variance all vehicles. Trace 6 is the hacked vehicle.

Fig. 11. CompRes results of the attack simulation 11 shown in Fig. 10c. (Dataset 3 Sim. attack 11; 60 frames; 50 cells; $p = 0.5$; $d = 0.1815$ [26]).

We have shown with CompRes that minimum and maximum boundaries must be considered for a dynamical system to be resilient. This is novel because system designers typically consider resilience as not going below minimum *quality of service* (QoS) but we did not find a definition that considers what happens if going beyond envisioned performance. This situation affects resilience because an overworking system is prematurely consuming its life expectancy.

In contrast to other resilience measurements, such as Ouyang and Dueñas-Osorio [19] our definition has three important advantages. First, it does not consider maximum performance as the $TP(t)$ of the system and therefore, maximum resilience is not obtained at 100% system output. Second, it provides a mechanism to express resilience without finding the optimal $TP(t)$. Third, it does not assume that a less resilient system should be closer to 0%. While undeniably a $P(t) = 0$ is not resilient because it is not working, in practise we do not need a non-working system to find out it is not usable. The resilience region also offers interesting benefits due to its relatively easy comprehension and predictable data consumption (e.g., Jabbar et al. [12] seems more data intensive). As a shortcoming, working with two metrics instead of one has its own challenges. In most cases the lower bound resilience will be the most important since it implies the system will stop performing, but in many safety cases the upper bound will be the one to consider.

There are some limitations in the presentation. One of them is caused by the way how we constructed the resilience and attack regions since we expand them by simple comparison to the previously calculated minimum and maximum. This is not a problem for the resilience region but could be misleading for the attack one since the overlap of these regions does not mean system failure. This happens because the attack region does not consider probability of an attack and a single value out of hundred simulations could significantly push boundaries. We therefore think that the attack region should be interpreted as a risk zone rather than imminent failure zone. Another limitation is that the attack region hides the too-good-to-be-true performance c. Basically, it was possible to get better than expected performance on some densities under attack. This counter-intuitive behaviour highlights the necessity of the maximum resilience boundary since a system performing better than expected could be under attack.

The resilience region methodology also has some interesting consequences when improving resilience, a process typically pursued by following the notional resilience profile [14]. This implies boosting the avoidance, absorption, recovering and/or adaptation to attacks. The first three characteristics are straightforward and typically covered by engineering disciplines, but the four one represents a major challenge for cyber-physical systems. For example, in UTNs, *avoid* is achieved by extensive planning and preparing; *absorb* is done by new materials and constructions techniques and *recover* by faster fixing or reconstruction methods. *Adaptability* is believed to be granted by using computers; however, we need to be realistic about what we can achieve with software.

Adaptation suggest that we should have systems capable of dealing with unexpected situations by implementing some form of artificial intelligence. However, if a resilient system needs to perform under strict minimum and maximum values, producing a more resilient system that adapts would require that those minimum and maximum values be modified. This is a problem because if the cyber-system is either over or under the original design performance limits, the software is going to classify the "adapted" behaviour as outside of the original resilience region and would try to avoid it. A computerised system will not be able to contradict itself and consequently will not increase its adaptability.

Nature solves adaptability by replacing the "component" (individual or full ecosystems) with a new mutated version, therefore adaptability becomes a characteristic of a specie rather than an individual. If we follow nature's approach, to create more resilient dynamical systems, including UTNs, we need to mimic this "evolutionary specie" idea in our cyber-physical designs. This in practise means not only to redefine the notional resilience profile to include *replace* as an adaptation mechanism but to design systems that balance the four characteristics. In UTNs, for example, engineers know that creating roads that last long time may involve very expensive materials; cheaper materials may not be as good but could be easier to replace and fix, therefore, they lean towards cheaper materials. We should consider this when designing resilient software to control cyber-physical systems.

8 Further Research

Future versions of this work could expand on different kind of deciders. This could allow detecting attacks on different kind of commodity flows (water, gas, electricity, etc.) using the presented resilience region methodology.

We also think that there could be better ways to generate the regions (resilience and attack) that could provide clearer risk indications (maybe with some probability coefficient).

Finally, the presented methodology could be seen as optimisation-oriented resilience. Seeing resilience as an optimisation process opens the door for exploring resilience using machine learning algorithms. We think this would be a very interesting area to expand the presented work since a more intelligent system is in essence a more resilient one.

9 Conclusions

In this article we presented a methodology to detect attacks on dynamical systems such as urban transport networks. These attacks are not necessarily evident but could be detected by analysing the expected performance behaviour of the system. Following the presented methodology, it is possible to study some non-evident factors that could improve the resilience of such systems.

We presented in this paper a way to model security and resilience on dynamical systems taking advantage of their stochastic behaviour. Being aware of this behaviour could be an advantage to safeguard systems.

The presented resilient region methodology is relatively straightforward. It is applicable to computerised models, such as a cellular automaton that could model a real dynamical system. This technique could be used to monitor real infrastructure and determine when a system with similar performance characteristics is under attack.

Additionally, we showed that a cellular automaton, such as the NaSCh simulator, could be used to study cyber-security. Even if they were not originally designed to do so. A significant amount of emphasis has been placed on attacks that involve decision making (decision attacks) and behaviour (link attacks). Notably, decision attacks could be crafted to be less evident. Cyber-physical systems, such as cyber UTNs, make possible these non-evident attacks.

Finally, we presented an interesting resilience definition that could be usable to other areas, since in essence covers the same functionality, it is quantifiable, and in general terms, could be sometimes easier to implement.

Acknowledgements. This work was funded by the Engineering and Physical Sciences Research Council (U.K.) through a DTP studentship (EP/R51309X/1)

References

1. Aponte, A., Moreno, J.A.: Cellular automata and its application to the modeling of vehicular traffic in the city of Caracas. In: El Yacoubi, S., Chopard, B., Bandini, S. (eds.) ACRI 2006. LNCS, vol. 4173, pp. 502–511. Springer, Heidelberg (2006). https://doi.org/10.1007/11861201_58
2. Barthélemy, M.: Spatial networks. Phys. Rep. **499**(1), 1–101 (2011). ISSN 0370–1573. https://doi.org/10.1016/j.physrep.2010.11.002
3. Bernot, M.: Optimal Transportation Networks: Models and Theory. Lecture notes in mathematics (Springer-Verlag); 1955. Springer, Berlin (2009). ISBN 9783540693154. https://doi.org/10.1007/978-3-540-69315-4
4. Bilal, K., Manzano, M., Erbad, A., Calle, E., Khan, S.U.: Robustness quantification of hierarchical complex networks under targeted failures. Comput. Electr. Eng. **72**, 112–124 (2018). ISSN 0045–7906. https://doi.org/10.1016/j.compeleceng.2018.09.008
5. Bishop, M., Carvalho, M., Ford, R., Mayron, L.M.: Resilience is more than availability. In: Proceedings of the 2011 New Security Paradigms Workshop, pp. 95–104 (2011)
6. Buldyrev, S.V., Parshani, R., Paul, G., Stanley, H.E., Havlin, S.: Catastrophic cascade of failures in interdependent networks. Nature **464**(7291), 1025–1028 (2010)
7. Chien, Y.H.: The optimal preventive-maintenance policy for a NHPBP repairable system under free-repair warranty. Reliab. Eng. Syst. Safety **188**, 444–453 (2019). ISSN 0951–8320. https://doi.org/10.1016/j.ress.2019.03.053
8. Greenburg, A.: Hackers remotely kill a jeep on the highway–with me in it. Wired 21st July 2015 (2015)
9. Heiser, G.: For safety's sake: we need a new hardware-software contract! IEEE Des. Test **35**(2), 27–30 (2018)
10. Heracleous, C., Kolios, P., Panayiotou, C.G., Ellinas, G., Polycarpou, M.M.: Hybrid systems modeling for critical infrastructures interdependency analysis. Reliab. Eng. Syst. Safety **165**, 89–101 (2017). ISSN 0951–8320. https://doi.org/10.1016/j.ress.2017.03.028

11. Hu, F., Yeung, C.H., Yang, S., Wang, W., Zeng, A.: Recovery of infrastructure networks after localised attacks. Sci. Rep. **6**(1), 24522 (2016). https://doi.org/10.1038/srep24522
12. Jabbar, A., Narra, H., Sterbenz, J.P.: An approach to quantifying resilience in mobile ad hoc networks. In: 2011 8th International Workshop on the Design of Reliable Communication Networks (DRCN), pp. 140–147. IEEE (2011)
13. Janić, M.: Modelling the resilience, friability and costs of an air transport network affected by a large-scale disruptive event. Transport. Res. Part A: Policy Pract. **71**, 1–16 (2015). ISSN 0965–8564. https://doi.org/10.1016/j.tra.2014.10.023
14. Kott, A., Linkov, I. (eds.): Cyber Resilience of Systems and Networks. Risk, Systems and Decisions, Springer, 1st ed. 2019. edn. ISBN 9783319774923. Springer, Cham (2019). https://doi.org/10.1007/978-3-319-77492-3
15. Kumar, S., Espling, U., Kumar, U.: Holistic procedure for rail maintenance in Sweden. Proceed. Instit. Mech. Eng. Part F: J. Rail Rapid Transit **222**(4), 331–344 (2008). https://doi.org/10.1243/09544097JRRT177
16. Myers-Smith, I.H., Trefry, S.A., Swarbrick, V.J.: Resilience: easy to use but hard to define. Ideas Ecol. Evol. **5**, 44–53 (2012)
17. Nagel, K., Schreckenberg, M.: A cellular automaton model for freeway traffic. J. De Phys. **I**(2), 2221–2229 (1992)
18. Ouyang, M.: Review on modeling and simulation of interdependent critical infrastructure systems. Reliab. Eng. Syst. Safety **121**, 43–60 (2014). ISSN 0951–8320. https://doi.org/10.1016/j.ress.2013.06.040
19. Ouyang, M., Dueñas-Osorio, L.: Time-dependent resilience assessment and improvement of urban infrastructure systems. Chaos (Woodbury, N.Y.) **22**(3), 033122 (2012). ISSN 1054–1500. https://doi.org/10.1063/1.4737204
20. Petersen, L., Lange, D., Theocharidou, M.: Who cares what it means? Practical reasons for using the word resilience with critical infrastructure operators. Reliab. Eng. Syst. Safety **199**, 106872 (2020). ISSN 0951–8320. https://doi.org/10.1016/j.ress.2020.106872
21. Pitt, S.M.: Learning lessons from the 2007 floods: an independent review: interim report. Cabinet Office (2007)
22. Pregnolato, M., Ford, A., Glenis, V., Wilkinson, S., Dawson, R.: Impact of climate change on disruption to urban transport networks from pluvial flooding. J. Infrastruct. Syst. **23**, 372 (2017). https://doi.org/10.1061/(ASCE)IS.1943-555X.0000372
23. Proag, V.: The concept of vulnerability and resilience. Procedia Econ. Finan. **18**, 369–376 (2014). ISSN 2212–5671. https://doi.org/10.1016/S2212-5671(14)00952-6. In: 4th International Conference on Building Resilience, Incorporating the 3rd Annual Conference of the ANDROID Disaster Resilience Network, 8th - 11th September 2014. Salford Quays, United Kingdom
24. RAIB: Derailment of a passenger train at Carmont, aberdeenshire - 12 august 2020. Tech. Rep., Rail Accident Investigation Branch (2020)
25. Rodríguez-Núñez, E., García-Palomares, J.C.: Measuring the vulnerability of public transport networks. J. Transport Geograp. **35**, 50–63 (2014). ISSN 0966–6923. https://doi.org/10.1016/j.jtrangeo.2014.01.008
26. Sanchez, D.: Compres dataset samples (2022). https://github.com/david-San/CompRes/tree/main/public_html/resources/datasets
27. Schadschneider, A.: Traffic flow: a statistical physics point of view. Physica A: Statist. Mech. Appl. **313**(1), 153–187 (2002). ISSN 0378–4371. https://doi.org/10.1016/S0378-4371(02)01036-1. Fundamental Problems in Statistical Physics

28. Treiber, M., Kesting, A.: Traffic flow dynamics. Springer, 1st edn. (2013). https://doi.org/10.1007/978-3-642-32460-4
29. Wang, X., Koç, Y., Derrible, S., Ahmad, S.N., Pino, W.J., Kooij, R.E.: Multi-criteria robustness analysis of metro networks. Physica A: Statist. Mech. Appl. **474**, 19–31 (2017). ISSN 0378–4371. https://doi.org/10.1016/j.physa.2017.01.072
30. Weick, K.E.: The vulnerable system: An analysis of the Tenerife air disaster. J. Manag. **16**(3), 571–593 (1990). https://doi.org/10.1177/014920639001600304
31. Zhou, X., Wu, C., Li, Y., Xi, L.: A preventive maintenance model for leased equipment subject to internal degradation and external shock damage. Reliab. Eng. Syst. Safety **154**, 1–7 (2016). ISSN 0951–8320. https://doi.org/10.1016/j.ress.2016.05.005

Analysis of the Battery Level in Complex Wireless Sensor Networks Using a Two Time Scales Second Order Fluid Model

Marco Gribaudo[1], Mauro Iacono[2(✉)], Daniele Manini[3], and Michele Mastroianni[4]

[1] Politecnico di Milano, via Ponzio 4/5, 20133 Milano, Italy
marco.gribaudo@polimi.it
[2] Università degli Studi della Campania "L. Vanvitelli", viale Lincoln 5, 81100 Caserta, Italy
mauro.iacono@unicampania.it
[3] Università degli Studi di Torino, Corso Svizzera 185, 10149 Torino, Italy
daniele.manini@unito.it
[4] Università degli Studi di Salerno, via Giovanni Paolo II 132, 84084 Fisciano, Italy
mmastroianni@unisa.it

Abstract. Battery operated Wireless Sensor Networks (WSNs) are currently one of the most important research areas related to applications: the possibility of running complex algorithms and off-load tasks using Fog and Edge computing techniques, as well as the ability of increasing the battery lifetime adopting energy harvesting, together with the communication capabilities offered by infrastructures such as 5G, are just some of the reasons for which this topic continues to be one of the most challenging and interesting. In this paper we focus on the problem of modeling the battery evolution of the devices, focusing on the issues created by the two time scales at which system evolves: tasks execution, sensor readings, and network operations occur at a small time scale, while energy harvesting and battery depletion runs on a much larger time frame. We propose a fluid model for the entire system, and we analyze it in two steps: we first focus on the small time scales to produce a simplified Second Order Fluid Model (SOFM), which later is used to reproduce the evolution at the large time frame. We analyze the considered models using discrete event simulation.

Keywords: Wireless Sensor Networks · Second Order Fluid Models · Simulation · Edge computing · Computing

1 Introduction

The diffusion of Internet of Things (IoT) devices and of specific wireless network technologies, such as ZigBee or LoRaWan, allows the implementation of cost-effective, distributed computing systems that may consist of a large number

M. Forshaw et al. (Eds.): PASM 2022, CCIS 1786, pp. 47–60, 2023.
https://doi.org/10.1007/978-3-031-44053-3_3

of independent nodes that are suitable to perform sensing and computing on environmental inputs in large open areas, such as campuses, forests, sea areas or volcanoes. IoT nodes may be used to remotely monitor wild, uninhabited areas in order to protect them from offenders, detect the passage of illegal traffic, study wildlife or prevent emergencies.

Market allows the design and implementation of battery-powered nodes that are equipped with sensors, cameras, a computing subsystem that can execute significant programmable tasks, low-power wireless networking and energy harvesting devices, according to the opportunities offered by the installation site or conditions. Harvesting technologies, that allow a node to extract energy from the surrounding, e.g. from vibrations if installed on moving objects as train goods carriages, from the sun if equipped with solar panels, from mechanical interactions if equipped with piezoelectric elements or analogous devices, from the electromagnetic field in which they are immersed, can mitigate the limitations caused by the installation in remote areas and partially or totally compensate the energy drained by operations, in case by applying a smart scheduling of node tasks.

The design process for these systems needs to be supported by proper modeling techniques, to represent the operating conditions, the evolution and the energy balance of a node, and performance evaluation tools, to check the actual matching of specifications in different potential operating conditions, including, in the case of interest for this paper, the potential coverage in terms of available operation time when the node and the IoT network are stressed by the targets that should be modeled in any moment of the operational life of the monitoring system.

As the variability and the potential intensity of the workload distributed over the IoT network increases the complexity of the evolution in the state space of the system, realistic predictions are needed to calibrate costs in the design and configuration of single nodes or the whole network, depending on the target areas, and the planning of rescue or maintenance missions.

In our previous works [2,4,5], we presented an IoT-based, energy harvesting, distributed monitoring system for extended, wild areas, capable of optimizing the lifespan of the network and the survivability of the overall monitoring capability by means of a task offloading mechanism and of supporting edge computing features, leveraging the onboard resources. In those papers, we explored system-oriented performance evaluation strategies to support the design cycle of the system and presented related analyses, including the influence of the environment on the system and of variable behaviors of different actors; in this paper we focus on the analysis of energy utilization in a single IoT node, by exploiting a two-levels hierarchical second order fluid model to model a node in detail.

This paper is organized as follows: after this introduction, related work is presented in Sect. 2; then, the WSN node is described in Sect. 3 and modeled in Sect. 4; the model is analyzed in Sect. 5; conclusions and future work close the paper.

2 Related Work

The problem of energy management and evaluation and its consequences on performance is relevant and literature is tackling the issue, also in the direction of green computing. We limit this section to some studies we found to be helpful in understanding the problem with relation to the purposes of our project, specially at the network level.

A simulation-based approach is used in [15], that deals with mobile edge computing. In this paper application-related issues are considered, in the context of energy-constrained IoT devices with limited power source. The target is the evaluation of expected performance of the applications, including low-latency applications that operate on high-resolution streams of video information and offloading features towards the cloud. The authors propose an algorithm that considers the impact on energy consumption of different strategies, including different parameters for visual information management and processing, to find trade-offs and help deciding about the kind and equipment of nodes, leveraging the ns-3 [3] network simulator.

Edge computing scenarios for IoT are also targeted in [11]. In this paper the authors present a specific simulation environment, built by extending the well-known CloudSim cloud systems simulator. The tool presented in the paper, IoTSim-Edge, leverages CloudSim core and libraries to include in the framework the means to simulate the features of edge devices and IoT devices, including different IoT networking protocols and their energy-related aspects and different protocols from the application layer. The paper also provides a synopsis of the parameters that model the added technologies, such as WiFi, LTE, Bluetooth, HTTP, XMPP and other protocols of different ISO/OSI levels.

Analytical methods have been used for evaluating energy management techniques in several kinds of wireless networks. An example is provided by Lyapunov-based optimization algorithms, used to minimize the overall use of energy: in [10] offloading is used to balance the available energy in the network in the case of mobile handheld devices by moving workloads to remote computing resources in a dynamic fashion, with the purpose of extending the lifetime of handelds. In fact, offloading is a popular technique that has become one of the characteristics of 5G mobile communication systems, providing low-cost resources by means of edge computing or cloud facilities integrated in the overall system: offloading is one of the characteristics of our solution as well, but, in this paper, it is considered as other normal tasks on a single node, as offloaded workloads are homogeneous with autochthonous workloads. Another example is provided in [16], that also deals, with larger attention to details related to network and node energy consumption aspects, with energy management in a multiuser Mobile Edge Computing system, including energy harvesting devices as well.

Game theory is also a tool applied by literature to such problems. Also in [7] offloading is considered as the main tool to balance energy consumption and enhance energy management features in a mobile edge computing architecture, considering from the structural point of view the game and deriving useful considerations about the possibility of a Nash equilibrium solution that considers the multiple users of the system, and analogous solutions are proposed in [13] and [12].

In our previous works related to this project, we relied onto simulation-based solutions for the evaluation of energy management techniques suitable for our IoT edge computing based monitoring system. In [4] we adopted an hybrid simulation technique to include also the effects of monitored targets of various categories and different dynamic behaviors, as they solicit in different ways the network and the nodes; we also proposed a DSL-based approach for modeling similar systems [2]. We also experimented, with special reference to offloading, a fuzzy-based approach [5]. An extended selection of references can be found in these papers.

In this paper we focus on a single node by exploiting a second-order fluid modeling technique similar to the one we applied in [1]: for the sake of space, we suggest the reader to refer to it for details and further related work.

3 The WSN Node

A node of the WSN described in Sect. 1 is configured with a computing unit, that consists of a CPU with its RAM and the controllers needed to interface to all hardware components, sensors, a LoRaWan network interface, a camera, a battery and a solar panel. The node runs software tasks by means of its CPU, that alternates intense computing phases to idle periods. When the CPU is computing, it switches to its high performance mode and absorbes 50 mA, while in the idle periods it goes into low consumption mode and absorbes 10 mA.

Some of the tasks are run on a regular basis, and form the baseline of the CPU activity. They manage ordinary operations, including basic sensing support and sensed data analysis, node management, self diagnosis and network coordination; other tasks involve network operations and the management of camera operations, including the basic processing of acquired videos.

Camera related tasks are triggered by camera activities, when an anomaly is detected by normal sensing operations and direct monitoring of the area is needed. In alarm mode, the camera is turned on for all needed time and tracks the target, absorbing 80 mA. The camera generates video information chunks that are processed by proper hardware in sequence, absorbing 50 mA. The processed video information is transmitted over the network to a remote control site for direct monitoring, that, in case, sends a team for intervention in case of offenders or other problems. This activity also causes additional CPU workload, launching other software tasks that are executed as soon as the CPU is found into idle mode, not to interfere with ordinary tasks.

The LoRaWan network interface is in charge of transmitting ordinary sensing information over the network, on a periodic basis, plus video information produced by the video processing hardware. The LoRaWan interface absorbes 100 mA when transmitting or receiving.

All subsystems are powered by the node battery, that has a capacity of 2 Ah. The node has harvesting capabilities, based on a solar panel that can provide a maximum of 0.5 W and 100 mA in optimal conditions.

4 Modeling the Node

The dynamics of the node may be described by a state space that has a discrete component, related to the instantaneous operating conditions of the CPU, the camera, the network interface and the transitions between them, and a continuous component, related to the instantaneous capacity of the battery, its usage over time as a result of the persistence of specific instantaneous operating conditions of the aforementioned subsystems, and the effect of the harvesting subsystem, that is continuously providing energy to the battery depending on the instantaneous sunlight exposure conditions of the solar panel. As the exposure conditions vary during a day and according to the weather conditions, their effects should be included into the model.

The node is characterized by operations that can be classified into two different time scales: the harvesting subsystem and the battery usage belong to the slow time scale subset, while all other subsystems belong to the fast time scale subset.

Due to the fact that the system is dominated by two different time scales, one related to internal node operations, including routine computing operations and sensing data transmission, and the other related to the events related to the external world, mainly the need for camera operations and the harvesting cycle by means of the solar panel, the model will be analyzed by means of a two-steps process: the model is partitioned into a low speed portion and a high speed portion, then the high speed portion is evaluated in isolation to obtain a synthetic description of its behavior to insert it into a modified version of the low speed portion. The model includes submodels that represent the behavior of single components that evolve between discrete conditions, such as the ones devoted to sensing, computing, transmissions, camera operations and related data processing in the high speed portion and to harvesting in the low speed portion, and the part that models the energy aspects of the system, exhibiting a continuous behavior with respect to battery utilization and acting as a bridge between subsystems playing the role of "battery users" and the harvesting subsystem.

4.1 The Model

The system has been modeled by means of a Fluid Stochastic Petri Net (FSPN) [8], shown in Fig. 1. Portions of the model that may be considered as submodels related to different components are enclosed into dashed contours.

Submodel Sc describes the behavior of the camera of the node. A token in place S_{OFF} represents the condition in which the camera is off, so it is not soliciting other subsystems of the node; a token in place S_{ON} represents the condition in which the camera is operating, and produces data for the node; the submodel is such that these two places are alternatively marked by a single token. Transition r_1 represents the average time in which the camera is on when the node is solicited by an external agent, while transition r_2 represents the average time between one activation and another[1]. When S_{ON} is marked, the camera generates data that must be processed by the video subsystem.

[1] This is a simplification of the actual solicitation model, suitable for the purposes of this paper: we considered more realistic scenarios in [2].

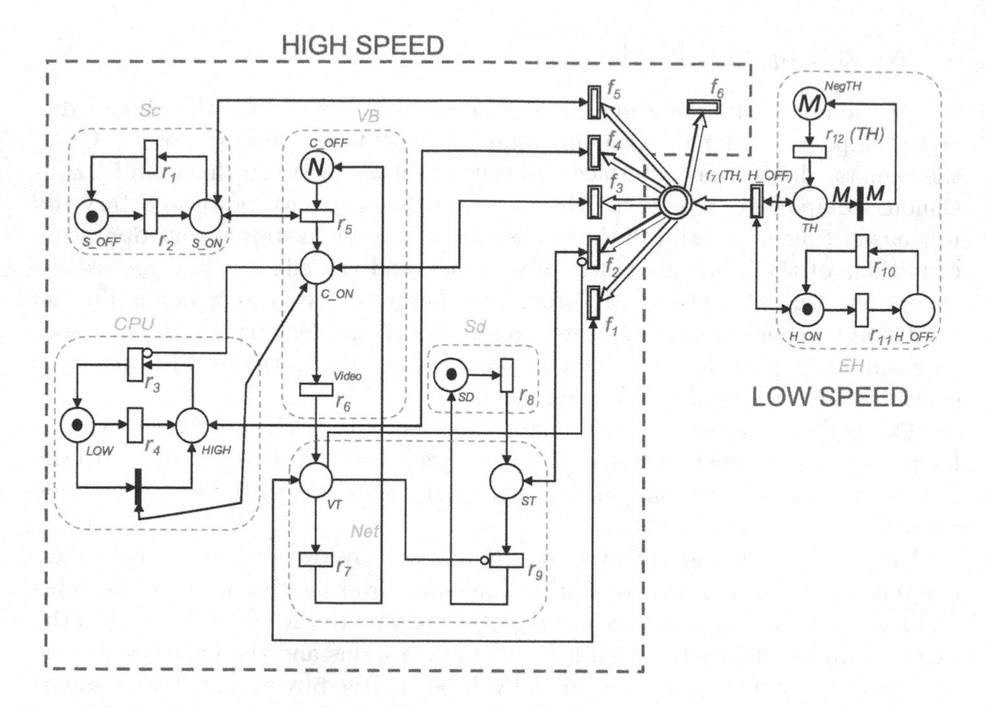

Fig. 1. The Fluid Stochastic Petri Net model of the WSN node.

Submodel VB describes the behavior of the video subsystem. Tokens in place C_{OFF} represent the available positions in the video buffer, that amount to N; a token in place C_{ON} represents the presence of data from the camera to be processed by the computing resources that are local to the video subsystem. Transition r_5 represents the average arrival time between two workloads generated by the camera, when active, while transition r_6 represents the average time needed by the local computing resources to process a workload unit. Each time the video subsystem has some workload to process, the CPU is also impacted by an additional workload with respect to its standard periodic workload, so, when C_{ON} is marked, the CPU processes it as soon as it finishes its current operation. When the video subsystem completes its operations on a camera workload, related information has to be transmitted by the network subsystem.

Submodel Net describes the behavior of the network subsystem, for what concerns data transmission. Tokens produced in places VT and ST represent data to be transmitted, respectively produced by camera operations and baseline periodic sensing operations. Camera operations produce larger volumes of data, that require longer transmissions with respect to data produced by baseline periodic sensing operations, so that the average transmission times are respectively represented by transitions r_7 and r_9. Camera operations are supposed to happen in support to critical situations, so that they get a higher priority in transmissions.

Baseline periodic sensing operations are represented by submodel *Sd*: transition r_8 represents the periodicity of the production of a related information unit to be transmitted over the network.

Submodel *CPU* describes the behavior of the main computing subsystem of the node. A token in place *LOW* represents the condition in which the CPU is in idle state, with minimal energy requirements; a token in place *HIGH* represents the condition in which the CPU is processing some workload; the submodel is such that these two places are alternatively marked by a single token. Transition r_3 represents the average processing time for a computing task, while transition r_4 represents the average time between one activation and another of the CPU according to the periodic tasks of which its baseline workload consists. When the video subsystem produces additional workload, related tasks solicit the main computing subsystem as soon as the CPU is in an idle state, producing an additional period in the condition represented by a token in the *HIGH* place, that depends on the duration of video subsystem operations.

Rates for the timed transitions are listed in Table 1.

Table 1. Transition, related meaning and average firing time.

Name	Meaning	Value
r_1	Active time for the camera	30 s
r_2	Time between camera activities	5 min
r_3	Periodic task compute time	0.4 s
r_4	Time between periodic tasks	12 s
r_5	Camera frame rate	0.25 s
r_6	GPU frame processing time	0.2 s
r_7	Video data transmission time	0.23 s
r_8	Time between periodic sensing transmissions	120 s
r_9	Sensing data transmission time	0.1 s
r_{10}	Duration of harvesting problems	6 h
r_{11}	Frequency of harvesting problems	2.75 days
r_{12}	Harvesting modulation time	(function)

As seen, some operational conditions of the presented subsystems cause battery drain. This is represented by the double arcs that connect the related places to fluid rates f_1–f_5. As all elements work with the same voltage, suitable for the battery, only currents are considered in the model. Currents required by the subsystem according to each condition of the systems are listed in Table 2. Fluid rate f_6 represents the baseline required by the system to stay on and perform basic sensing operations, including CPU needs in idle state, so that the value used for f_4 is set to the additional need in the high performance state of the CPU only (40 mA out of the 50 mA needed in this state). All values for fluid rates are negative, as they represent battery utilization.

Table 2. Places, related conditions, required currents and fluid rates.

Name	Meaning	Value	Fluid rate
S_OFF	Camera off	negligible	–
S_ON	Camera on	80 mA	f_5
LOW	CPU idle state	10 mA	included in baseline
$HIGH$	CPU active state	40 mA	f_4 plus baseline
C_OFF	Frame buffer available slots	negligible	–
C_ON	Frames being processed	50 mA	f_3
VT	Frames transmission	100 mA	f_1
SD	Normal sensing activity	1 mA	included in baseline
ST	Sensing data transmission	100 mA	f_2
–	Baseline	11 mA	f_6

Submodel EH in Fig. 1 represents the energy harvesting process, and has been designed to be easily customizable, to be adapted to consider different exposure and weather conditions. Submodel EH consists of two sections: the upper one manages the different phases of exposure during the 24 h, while the lower one manages the effects of bad weather conditions, that make the solar panel ineffective. The M total tokens marking the upper part regulate the phases of the day: marking of place TH represents the current phase. When M tokens are in TH, the cycle is terminated and all tokens are reset in place $NegTH$. In order to allow a different duration for each phase, the rate of transition r_{12} depends on the number of tokens in TH (that is, the current phase). The number of tokens in TH also influences the rate of fluid rate f_7 (as indicated by the control arc between the two model elements), representing the current generated by the panel according to the phase and the weather conditions. In fact, the lower part of EH modulates the production of the solar panel: when place H_{ON} is marked, weather conditions are optimal; when place H_{OFF}, weather conditions are suboptimal for energy harvesting. Optimal and suboptimal conditions alternate in accordance with the rates of transitions r_{10} and r_{11}[2]. Table 3 describes the actual values assumed by rates of fluid rate f_7 and transition r_{12}: the system is set to 4 phases, namely night, sunrise, daytime and sunset, that respectively last an average of 8, 4, 8 and 4 h and the panel delivers respectively 0, 50, 100, 50 mA in optimal weather conditions and 60% of those values in suboptimal conditions.

[2] Those rates must be set accordingly with the actual seasonal local weather phenomena: this is a simple modelization, but a similar approach as the upper part may be used to represent a more complex variability pattern. The effect of a token in the H_{OFF} place can alternatively represent both a lower efficiency or a deactivation of the solar panel, as needed by the real conditions of the actual system.

Table 3. Provided currents (mean and standard deviation) in different phases of the day and with different weather conditions.

Phase	Tokens in TH	$f_7(TH,1)$	$s_7(TH,1)$	$f_7(TH,0)$	$s_7(TH,0)$	$r_{12}(TH)$
Night	0	0	0	0	0	8 h
Sunrise	1	50 mA	10 mA	30 mA	6 mA	4 h
Daytime	2	100 mA	20 mA	60 mA	12 mA	8 h
Sunset	3	50 mA	10 mA	30 mA	6 mA	4 h

5 Analysis

We start focusing only on the part of the model contained in the `HIGH SPEED` frame of Fig. 1. In particular, we consider it as a reward model, and derive both its infinitesimal generator matrix $\mathbf{Q}$ and reward matrix $\mathbf{R}$. Elements r_{ii} of $\mathbf{R}$ accounts for the fluid transitions, as usual in FSPN analysis.

We then compute the steady state solution $\pi = |\pi_i|$ of the underlying CTMC (with $\mathbf{Q} = 0$ and $\sum \pi_i = 1$), and use it to derive the average power consumption:

$$\bar{r} = \sum_i \pi_i r_{ii} \tag{1}$$

We then create a new reward model, that starts at time $t = 0$ with $\pi(0) = \pi$, and the rate matrix of which is $\tilde{\mathbf{R}} = \mathbf{R} - \bar{r}\mathbf{I}$. This special reward model, due to this construction, has the property of having a constant average reward level of 0: in this way we can focus only on the analysis of its higher moments, and in particular on its second moment, that thanks to the zero mean, also corresponds to the variance of the process. Although efficient analytical techniques exist for computing the moments of the accumulated reward for this type of process (see for example [14]), they cannot be easily applied in this case due to the non-negligible size of the state space of the underlying CTMC, and the large temporal horizon. We thus perform discrete event simulation, generating traces of the accumulated reward. For example, Fig. 2a shows 25 traces of the accumulated reward at $t = 1$ h. We generate 62500 traces, grouped into 250 runs of 250 repetitions each, and use them to compute the second moment as function of time. Figure 2b shows the temporal evolution of the variance of the reward level, divided by the time, with its 95% confidence intervals. This process rapidly stabilizes at the value that can be used as the variance in Gaussian approximation to the distribution of the accumulated reward.

Figure 3 compares the real distribution with the Gaussian approximation in the considered scenario. As it can be seen, the real distribution is characterized by a finite support, and it is not symmetrical with a little skewness toward the negative values. Nevertheless, the Gaussian approximation provides a reasonable fitting to the curve.

In additions to the parameters shown in Tables 2 and 1, we set the buffer size for the video packet to $N = 5$. From the study of the model with the fast dynamic

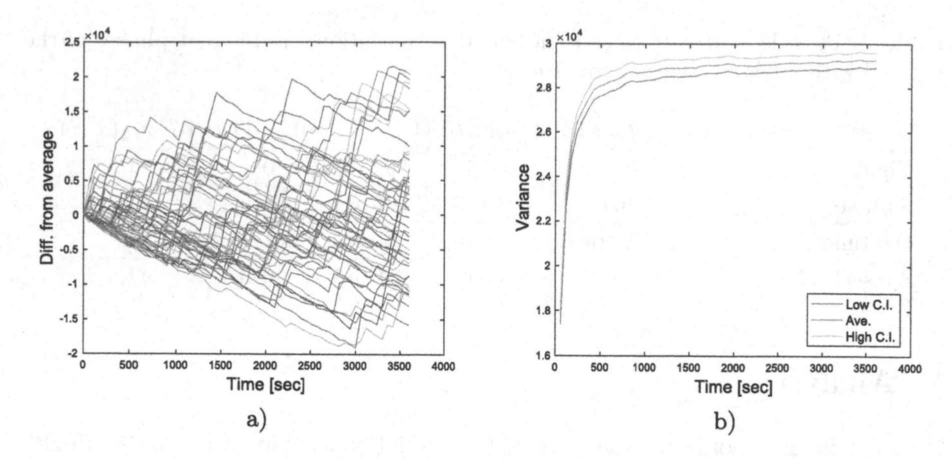

Fig. 2. Evolution of the power consumption of the fast component of the system: a) 50 traces of the difference from the average, b) evolution of the variance divided by the time.

we obtain an average power consumption of $\mu_1 = 183.29\,\text{mA}$, and a variance in the 95% confidence interval $[2.8912, 2.9607]\,\text{mA}^2$; we then set $\sigma_1 = 171.05\,\text{mA}$.

We use such parameters to study only the slow component of the system, using the Second Order FSPN shown in Fig. 4, where the Second Order fluid transition completely replace the fast component of the model. The initial marking of the considered fluid place corresponds to the capacity of the battery, which

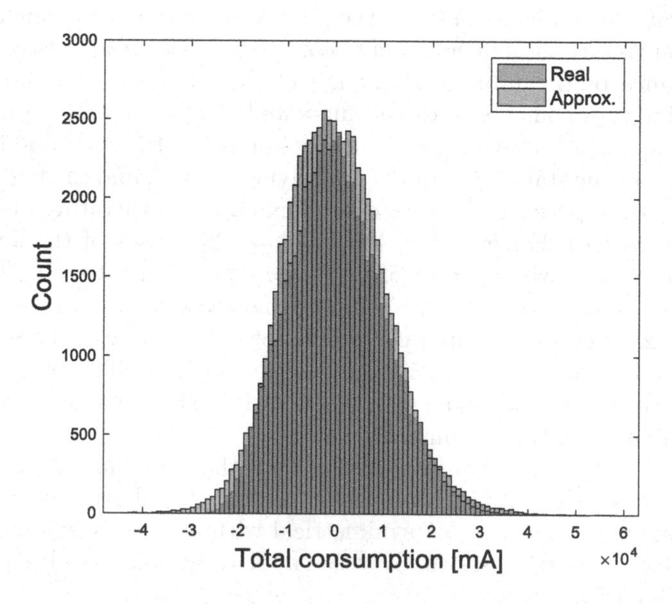

Fig. 3. Distribution of the difference form the average energy consumption and its Gaussian approximation at 1 h.

also corresponds to the upper bound for the continuous variable. But upper and lower boundary are characterized by a reflecting barrier: the system can leave from the empty state due to energy harvesting. Moreover, the battery cannot exceed its maximum capacity. For sake of simplicity, we consider a linear battery models: the framework can however easily support mor realistic battery models, following for example the technique used in [6]. We analyze the model using discrete event simulation, adopting in particular the techniques proposed in [9] to handle the boundaries. Due to its compactness, 200000 runs can be executed in about 1 min on a 2016 MacBook Pro laptop. Although confidence intervals are computed, the large number of runs make them indistinguishable from the mean in a conventional plot: for this reason they will not be shown.

We study the effect of three different battery sizes, namely 2000mAh, 4400 mAh, and 10000 mAh, as well as the effect of no energy harvesting, and of using three different sizes of solar panels: 0.5 W, 1 W and 2 W. Figure 5 shows 20 traces of the evolution of the battery level, for two different configurations: the ones with the minimum and maximum combination of battery capacity and solar panel size. While the first has the battery almost always depleted, and mainly depends on the energy harvested, the second shows a more interesting evolution determined by the events being monitored, and on the sate of energy harvesting.

Figure 6a shows the evolution of the average energy level with a 4400 mAh battery for different solar panel sizes. While with no harvesting the system dies in less than one day, energy harvesting increases the level and the duration of the system. In particular Fig. 6b shows the average battery level after 15 days from the start-up of the system. For solar panels up to 1 W, the size of the battery has practically no influence, and the system rapidly consumes all the energy available. Otherwise, with a 2 W solar panel, the battery size starts having an influence to the average level that can be found after two weeks.

To understand the implications of the battery depletion on the functionality of the system, Fig. 7 shows the probability of having the battery level at its lower or upper bound. The lower bound probability shown in Fig. 7a represents the probability that the system is not functioning due to unavailable energy. As it can see, when the harvesting procedure is not capable of providing enough energy, the probability of not being operational, which also corresponds to the probability of missing an important event, is almost independent on the battery level, and it is around 30% and 45%. When energy harvesting is sufficiently high, the missing event probability reaches acceptable levels: in this case a larger battery helps in further extending the operational state of the system and reducing the probability of not being operational. It is also interesting the probability of having a full battery, as shown in Fig. 7b. Only when harvesting is sufficiently high, this event can be effectively triggered. The interesting thing is that a battery with a smaller capacity, has a higher probability of reaching its full state.

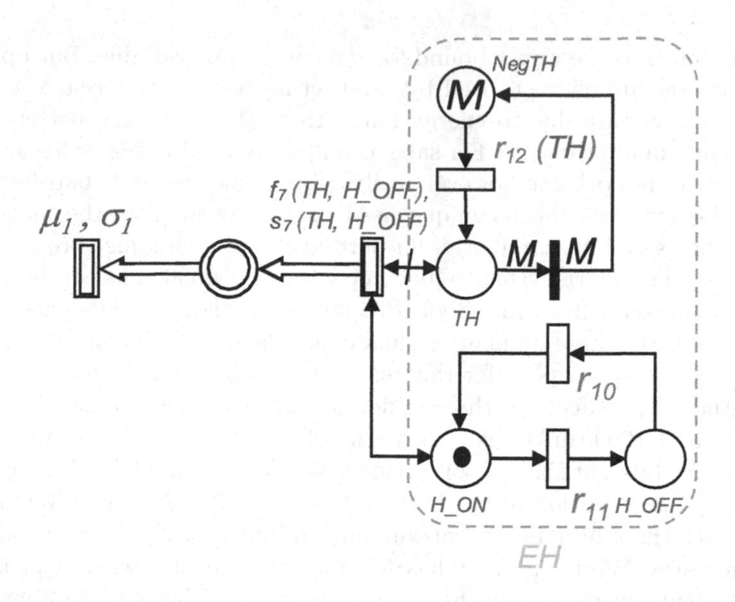

Fig. 4. Second Order FSPN of the slow dynamic of the system.

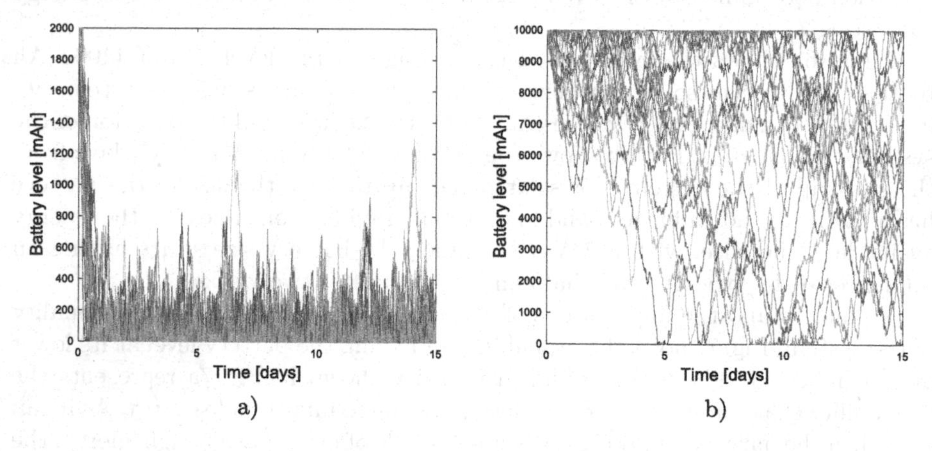

Fig. 5. Traces of the evolution of the battery level: a) Battery 2000 mAh, solar panel 0.5 W b) Battery 10000 mAh, solar panel 2 W.

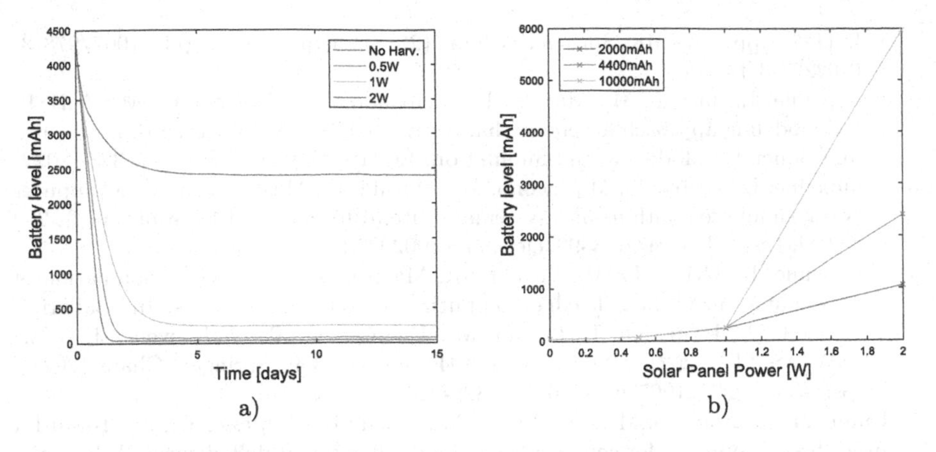

Fig. 6. Average battery level: a) Transient for the 4400mAh battery b) At 15 days for all the considered scenario.

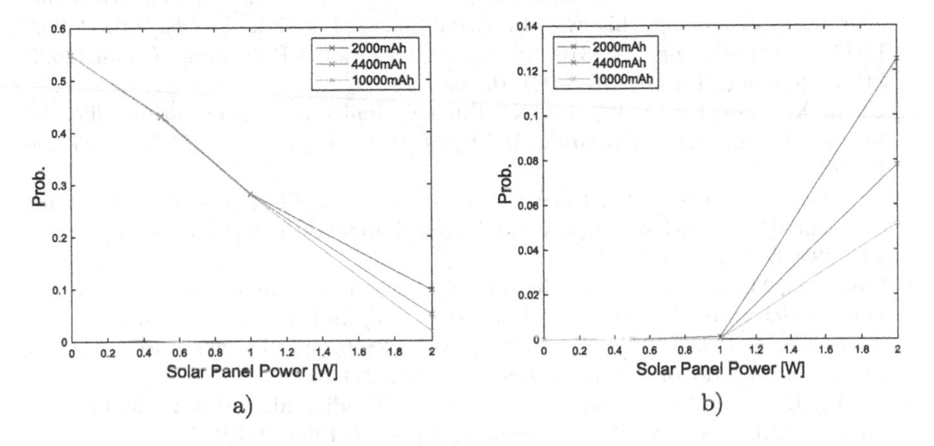

Fig. 7. Battery level at 15 days: a) Empty probability b) Full probability.

6 Conclusion and Future Work

Acknowledgements. This work has been partially funded by the internal competitive funding program "VALERE: VAnviteLli pEr la RicErca" of Università degli Studi della Campania "Luigi Vanvitelli" and is part of the research activities developed within the project PON "Ricerca e Innovazione" 2014–2020, action IV.6 "Contratti di ricerca su tematiche Green", issued by Italian Ministry of University and Research.

References

1. Barbierato, E., Gribaudo, M., Iacono, M., Piazzolla, P.: Second order fluid performance evaluation models for interactive 3D multimedia streaming. In: Bakhshi, R., Ballarini, P., Barbot, B., Castel-Taleb, H., Remke, A. (eds.) EPEW 2018. LNCS,

vol. 11178, pp. 205–218. Springer, Cham (2018). https://doi.org/10.1007/978-3-030-02227-3_14

2. Campanile, L., Iacono, M., Marulli, F., Gribaudo, M., Mastroianni, M.: A DSL-based modeling approach for energy harvesting IoT/WSN. In: Proceedings - European Council for Modelling and Simulation, ECMS, May 2022, pp. 317–323 (2022)
3. Campanile, L., Gribaudo, M., Iacono, M., Marulli, F., Mastroianni, M.: Computer network simulation with ns-3: a systematic literature review. Electronics **9**(2), 272 (2020). https://doi.org/10.3390/electronics9020272
4. Campanile, L., Gribaudo, M., Iacono, M., Mastroianni, M.: Hybrid simulation of energy management in IoT edge computing surveillance systems. In: Ballarini, P., Castel, H., Dimitriou, I., Iacono, M., Phung-Duc, T., Walraevens, J. (eds.) EPEW/ASMTA -2021. LNCS, vol. 13104, pp. 345–359. Springer, Cham (2021). https://doi.org/10.1007/978-3-030-91825-5_21
5. Campanile, L., Iacono, M., Marulli, F., Mastroianni, M., Mazzocca, N.: Toward a fuzzy-based approach for computational load offloading of IoT devices. J. Univers. Comput. Sci. **26**(11), 1455–1474 (2020)
6. Cerotti, D., Mancini, S., Gribaudo, M., Bobbio, A.: Analysis of an electric vehicle charging system along a highway. In: Ábrahám, E., Paolieri, M. (eds.) QEST 2022. LNCS, vol. 13479, pp. 298–316. Springer International Publishing, Cham (2022). https://doi.org/10.1007/978-3-031-16336-4_15
7. Chen, X., Jiao, L., Li, W., Fu, X.: Efficient multi-user computation offloading for mobile-edge cloud computing. IEEE/ACM Trans. Network. **24**(5), 2795–2808 (2016)
8. Gribaudo, M., Sereno, M., Horváth, A., Bobbio, A.: Fluid stochastic Petri nets augmented with flush-out arcs: modelling and analysis. Discret. Event Dyn. Syst. **11**(1/2), 97–117 (2001)
9. Gribaudo, M., Iacono, M., Manini, D.: Simulation of N-dimensional second-order fluid models with different absorbing, reflecting and mixed barriers. In: Abate, A., Marin, A. (eds.) QEST 2021. LNCS, vol. 12846, pp. 276–292. Springer, Cham (2021). https://doi.org/10.1007/978-3-030-85172-9_15
10. Huang, D., Wang, P., Niyato, D.: A dynamic offloading algorithm for mobile computing. IEEE Trans. Wirel. Commun. **11**(6), 1991–1995 (2012)
11. Jha, D.N., et al.: IoTsim-edge: a simulation framework for modeling the behavior of internet of things and edge computing environments. Softw. Pract. Exp. **50**(6), 844–867 (2020)
12. Liu, L., Chang, Z., Guo, X.: Socially aware dynamic computation offloading scheme for fog computing system with energy harvesting devices. IEEE Internet Things J. **5**(3), 1869–1879 (2018)
13. Shah-Mansouri, H., Wong, V.W.S.: Hierarchical fog-cloud computing for IoT systems: a computation offloading game. IEEE Internet Things J. **5**(4), 3246–3257 (2018)
14. Telek, M., Rácz, S.: Numerical analysis of large Markov reward models. Perform. Eval. **36–37**(1–4), 95–114 (1999)
15. Trinh, H., et al.: Energy-aware mobile edge computing and routing for low-latency visual data processing. IEEE Trans. Multimedia **20**(10), 2562–2577 (2018)
16. Zhang, G., Chen, Y., Shen, Z., Wang, L.: Distributed energy management for multiuser mobile-edge computing systems with energy harvesting devices and QOS constraints. IEEE Internet Things J. **6**(3), 4035–4048 (2019)

To Confine or Not to Confine: A Mean Field Game Analysis of the End of an Epidemic

Gontzal Sagastabeitia[1(✉)], Josu Doncel[1], and Nicolas Gast[2]

[1] University of the Basque Country, UPV/EHU, Leioa, Spain
gsagastabeitia001@ikasle.ehu.eus
[2] Univ. Grenoble Alpes, Inria, CNRS, Grenoble INP, LIG, Grenoble, France

Abstract. We analyze a mean field game where the players dynamics follow the SIR model. The players are the members of the population and the strategy consists in choosing the probability of being exposed to the infection, i.e., its confinement level. The goal of each player is to minimize the sum of the confinement cost, which is linear and decreasing on its strategy, and a cost of infection per unit time. We formulate this problem as a mean field game and we investigate the structure of a mean field equilibrium. We study the behavior of agents during the end of the epidemic, where the proportion of infected population is decreasing. Our main results show that: (a) when the cost of infection is low, a mean field equilibrium consists of never getting confined, i.e., the probability of being exposed to the infection is always one and (b) when the cost of infection is large, a mean field equilibrium consists of being confined at the beginning and, after a given time, being exposed to the infection with probability one.

Keywords: Mean field game · SIR model · Confinement

1 Introduction

The situation derived from COVID19 disease has put in evidence the need of carrying out research in the epidemic field. The most important epidemic model that has been investigated in the literature is based on the SIR model. In the SIR model, it is considered that each member of the population belongs to one of the following states: susceptible (S), infected (I) or recovered (R). It has been first studied in [13] and we refer to [1,5] for books presenting the large literature of the SIR model.

Mean field games study the rational behavior of an infinite number of players. They were introduced recently by Jean-Michel Lasry and Pierre Louis Lions in [16–18] and Minyi Huang et al. in [11]. Two important assumptions are made in mean field games: (a) the players are indistinguishable, i.e., one can only observe the number of objects in each state, and (b) as the number of players is

M. Forshaw et al. (Eds.): PASM 2022, CCIS 1786, pp. 61–80, 2023.
https://doi.org/10.1007/978-3-031-44053-3_4

infinite, the decisions of an individual player do not affect the dynamics of the whole population. These assumptions lead to a simplification of the computation of the equilibrium compared with the computation of the Nash equilibrium of games with a finite number of players, which is known to be a PPAD-complete problem[1] [4]. As a consequence, there has been in the last years a huge literature analyzing mean field games in a wide range of applications in macro-economic models [8], in autonomous vehicles [10], security of communications [14] and traffic modeling [2], to mention a few.

The SIR model and its variations have been also studied from the perspective of mean field games. Some works have analyzed mean field games where the action consists of the vaccinations policies, for instance [15] and its extension with births and deaths [12], and also [7]. In our work we do not consider vaccinations, but the confinement level, or in other words, how players choose to be exposed to the infection. We remark that there are some recent articles that have studied the effect of confinements in the population using mean field games [3,19]. In [3], players can choose a contact rate when they are infected or recovered, while in our work only the susceptible population controls their contact rate. Besides, they consider continuous time and a non-linear cost function, unlike we do. The authors in [19] also consider continuous time and that all players can choose the contact rate with the population. They also consider more variables, such as asymptomatic infected players and the population's age. The main difference between [3,19] and our work, besides the use of discrete time, is that we focus in the end of the epidemic, and that they do not show the existence of the mean field equilibrium nor study its structure. Another related work is [9] where they consider simultaneously the space-time evolution of the epidemics and of the human capital and focus on the benefits of formulating a mean field game.

In this work, we consider that there is an infinite number of players that can decide the probability of being exposed to the infection, i.e., the confinement level. We assume that there is a cost of confinement which is linear and decreasing on the strategy of the players and a cost of being infected per unit time. We also focus on a regime of the end of the epidemic in which the proportion of the infected population decreases over time. In this context, we formulate a mean field game and we show that the solution of this game, i.e., the mean field equilibrium, at the penultimate time step consists of being exposed to the infection with probability one. As a consequence, two strategies are considered in the following: (a) constant, which means that the probability of being exposed to the infection is always one, that is, that rational players are never confined and (b) one jump, which means the players are confined (i.e., not exposed to the infection), at the beginning and, from a given time, the probability of being exposed to the infection is one.

The main contributions of this article are summarized as follows:

[1] PPAD stands for "polynomial parity arguments on directed graphs". It is a complexity class that is a subclass of NP and is believed to be strictly greater than P.

- We establish sufficient conditions for the existence of a mean field equilibrium that is constant, i.e., when the mean field equilibrium consists of being exposed to the infection with probability one always (see (COND-CONST)). This condition means that, when the cost of infection is small, there exists a mean field equilibrium where players are never confined.
- We establish sufficient conditions for the existence of a mean field equilibrium that is a strategy with one jump (see (COND1-JUMP) and (COND2-JUMP)). This condition means that, when the cost of infection is large, there exists a mean field equilibrium where players are confined at the beginning and, from a given time, are completely exposed to the infection.

Finally, we discuss the numerical experiments we have carried out to analyze the structure of a mean field equilibrium when the aforementioned conditions do not hold. We conclude that when (COND1-JUMP) does not hold the mean field equilibrium with one jump does not seem to exist, but it might exist even though (COND2-JUMP) does not hold.

The rest of the article is organized as follows. In Sect. 2 we describe the model we study in this article and we formulate the mean field game under analysis. In Sect. 3 we present some preliminary results regarding the mean-field game. In Sect. 4 we present our results regarding the existence of a constant mean field equilibrium and in Sect. 5 we explain our result about the existence of a mean field equilibrium. In Sect. 6 we discuss the existence of a mean field equilibrium out of the conditions of our main results. In Sect. 7 we present the main conclusions of our work as well as the future research directions.

2 Model Description

2.1 Notation

We consider a population of homogeneous players that evolve in discrete time from 0 to T. The players are in one of the following three states: susceptible (S), infected (I) or recovered (R). We denote by $m_S(t)$, $m_I(t)$ and $m_R(t)$ the proportion of the population that is in each state.

The dynamics of one player is described as follows. A player encounters other players in a time slot with probability γ. If a player is susceptible and encounters an infected player, then it becomes infected. An infected player recovers in the next time slot with probability ρ. Once a player is recovered, its state does not change. We also consider that a susceptible player can be protected from the infection by choosing the strategy π. A strategy π is a function from $\{0, 1, \dots, T\}$ to $[0, 1]$ and $\pi(t)$ is the probability that a susceptible player at time slot t is exposed to the infection. When $\pi(t) = 0$, the players are confined, or in other words, they are completely protected from the infection at time t; on the other hand, when $\pi(t) = 1$, they are completely exposed to the infection and, therefore, they can be infected if they encounter an infected player. The Markovian behavior of a player is represented in Fig. 1.

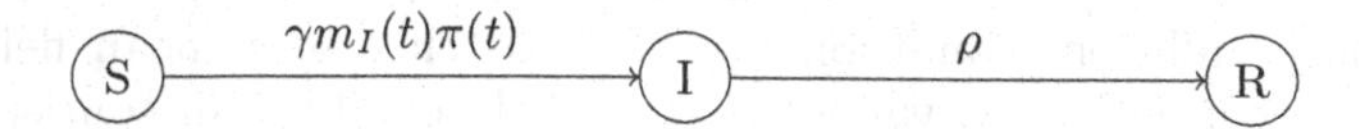

Fig. 1. The dynamics of a player in the epidemic model. An player has three possible states: S (susceptible), I (infected) and R (recovered).

We are interested in the analysis of this epidemic model with an infinite number of players. In this case, the dynamics of the population is given by the Kolmogorov Equation that takes the following form

$$\begin{cases} m_S(t+1) = m_S(t) - \gamma m_S(t) m_I(t)\pi(t) \\ m_I(t+1) = m_I(t) + \gamma m_S(t) m_I(t)\pi(t) - \rho m_I(t) \\ m_R(t+1) = m_R(t) + \rho m_I(t). \end{cases} \tag{1}$$

Let us make the following assumption.

Assumption 1. (m_I decreasing). *We assume that* $m_S(0)\gamma < \rho$.

We know from (1) that $m_S(t+1) \leq m_S(t)$ for all $t = 0, 1, \dots, T-1$, i.e., the proportion of the susceptible population is non-increasing with t. Therefore, from the above assumption, it follows that the proportion of infected population decreases with t:

$$\begin{aligned} m_I(t+1) &= m_I(t) + \gamma m_S(t) m_I(t)\pi(t) - \rho m_I(t) \\ &= m_I(t)(1 + \gamma m_S(t)\pi(t) - \rho) \\ &\leq m_I(t)(1 + \gamma m_S(0)\pi(t) - \rho) \\ &\leq m_I(t)(1 + \gamma m_S(0) - \rho) \\ &< m_I(t), \end{aligned}$$

where in the first inequality we use that $m_S(t) \leq m_S(0)$, in the second inequality that $\pi(t) \in [0,1]$ and in the last one the property of Assumption 1. In Sect. 6, we discuss the difficulties on the analysis of the formulated mean field game when Assumption 1 does not hold.

2.2 Mean Field Game Formulation

We focus on a particular player, that we call Player 0. As we consider a mean field game model, the dynamics of the global population are not affected by Player 0 alone and is driven by Eq. (1). Player 0 chooses her confinement strategy π^0, where $\pi^0(t) \in [0,1]$ for all $t = 0, 1, \dots, T$. We consider that, when $\pi^0(t) = 0$, Player 0 gets confined and therefore, it cannot be infected. But, when $\pi^0(t) = 1$, Player 0 is completely exposed to the infection. The probability that Player 0 is in a given state depends not only on π^0, but also on $m(t)$, the population distribution (which depends on $\pi(t)$).

Let us make the following assumption regarding the cost of a player.

Assumption 2. (Linear confinement cost). *We assume that when a player chooses strategy $\pi^0(t)$, its confinement cost is $c_L - \pi^0(t)$ at time t, where $c_L \geq 1$.*

Therefore the confinement cost of Player 0 is linear with respect to its confinement strategy, and the total cost incurred by the player is the sum of two costs: the confinement cost, which is presented in Assumption 2; and the infection cost, a constant $c_I > 0$ per time unit if the player is infected.

Let $x_i^{\pi^0,\pi}(t)$ be the probability that Player 0 is in state i at time t, where $i \in \{S, I, R\}$. The quantities $x_i^{\pi^0,\pi}(t)$ satisfy the following system of equations:

$$\begin{cases} x_S^{\pi^0,\pi}(t+1) = x_S^{\pi^0,\pi}(t) - \gamma x_S^{\pi^0,\pi}(t) m_I(t)\pi(t) \\ x_I^{\pi^0,\pi}(t+1) = x_I^{\pi^0,\pi}(t) + \gamma x_S^{\pi^0,\pi}(t) m_I(t)\pi(t) - \rho x_I^{\pi^0,\pi}(t) \\ x_R^{\pi^0,\pi}(t+1) = x_R^{\pi^0,\pi}(t) + \rho x_I^{\pi^0,\pi}(t). \end{cases}$$

Note that the above equation is similar to Eq. (1) except that it is linear in x whereas Eq. (1) is not linear in m.

The expected individual cost of Player 0 is:

$$\sum_{t=0}^{T} \left[x_S^{\pi^0,\pi}(t) f(\pi^0(t)) + c_I x_I^{\pi^0,\pi}(t) \right],$$

where $f(a) = c_L - a$ represents the cost of confinement in a time unit.

We call the *best response to* π, and denote it by $\mathrm{BR}(\pi)$, the set of confinement strategies that minimize the expected cost of Player 0 for a given population strategy π, that is,

$$BR(\pi) = \arg\min_{\pi^0} \sum_{t=0}^{T} \left[x_S^{\pi^0,\pi}(t) f(\pi^0(t)) + c_I x_I^{\pi^0,\pi}(t) \right], \tag{2}$$

which is a non-empty set by compacity of the strategy space.

We define a mean field equilibrium as a fixed point of the best-response function:

Definition 1 (Symmetric Mean Field Equilibrium). *The strategy π^{MFE} is a symmetric mean field equilibrium if and only if*

$$\pi^{MFE} \in BR(\pi^{MFE}).$$

This is the classical definition of an equilibrium in a mean field game. The rationale behind this definition is that in a homogeneous population, each player's best-response is the same as that of Player 0. This means that, for a given confinement strategy of the population π, any player of the population chooses the strategy $BR(\pi)$. As in classical games, a mean field equilibrium is a situation where no player has incentive to deviate unilaterally from the selected confinement strategy.

Remark 1. This model is a particular case of the mean field games studied in [6] and therefore, the existence of a mean field equilibrium follows directly. In this work, we go beyond this existence result and our goal is to characterize the structure of the solution of the formulated mean field game.

2.3 Discussion and Limits of Our Assumptions

We have already presented the two main assumptions made in this paper, Assumption 1 and 2, which are not found in most papers on this subject.

Assumption 1 means that we are only studying the end of the epidemic rather than its spread. Therefore, our results focus on strategies for lifting restrictions, rather than on preemptive measures. This assumption makes it easier to find the mean-field equilibrium, and provides more straightforward responses from the players. The generalisation of this work to the whole spread of the epidemic is left for future work.

On the other hand, the main effect Assumption 2 has on our results is the binary choice of players. The linear cost means that when minimizing costs, the best response of players will either be full lockdown or no measures at all. Therefore, instead of having a smooth best response, as in most papers that study similar models, the players' best response will not be continuous. This limitation makes the implementation of a strategy more clear, as the decisions are reduced to two options; but this lack of smoothness makes finding the mean-field equilibrium more difficult.

3 Preliminary Results

We focus on the best-response to π of Player 0. We know that the optimal cost and the best-response verify the following Bellmann equations: for $t = 0, 1, \ldots, T-1$,

$$\begin{aligned} V_S^*(t) &= \min_{\pi^0(t)\in[0,1]} \left(f(\pi^0(t)) + (1-\gamma m_I(t)\pi^0(t))V_S^*(t+1)\right. \\ &\qquad \left. +\gamma m_I(t)\pi^0(t)V_I^*(t+1)\right) \\ V_I^*(t) &= c_I + (1-\rho)V_I^*(t+1), \\ BR(\pi)(t) &= \operatorname*{arg\,min}_{\pi^0(t)\in[0,1]} \left(f(\pi^0(t)) + (1-\gamma m_I(t)\pi^0(t))V_S^*(t+1)\right. \\ &\qquad \left. +\gamma m_I(t)\pi^0(t)V_I^*(t+1)\right), \end{aligned}$$

with $V_S^*(T) = V_I^*(T) = 0$.

Remark 2. The best response of Player 0 for strategy π $BR(\pi)$ is the product of all the best responses at time t $BR(\pi)(t)$ over $t = 0, \ldots, T-1$. I.e.

$$BR(\pi) = \{\pi^0 | \pi^0(t) \in BR(\pi)(t),\ \forall t = 0, \ldots, T-1\}$$

We first note that, since $V_S^*(T) = V_I^*(T) = 0$, and from the above Bellmann equations, the best response at time $T-1$ is equal to 1, i.e., the best strategy of Player 0 at the penultimate time step is always one. Therefore, throughout the article, when we say that a strategy (the best-response or the mean field equilibrium) is constant, we mean that it is always one (see Sect. 4). Likewise, when we say that a strategy has one jump, we mean that there exists a value t_0

such that $0 \leq t_0 < T - 1$ and that the considered strategy is zero from 0 to t_0 and one from $t_0 + 1$ to T (see Sect. 5).

We remark that $V_S^*(t)$ is the value for $\pi^0(t)$ that minimizes the following function:

$$c_L - \pi^0(t) + (1 - \gamma m_I(t)\pi^0(t))V_S^*(t+1) + \gamma m_I(t)\pi^0(t)V_I^*(t+1).$$

The derivative with respect to $\pi^0(t)$ of the above expression is

$$-1 + \gamma m_I(t)(V_I^*(t+1) - V_S^*(t+1)).$$

This means that the derivative of $V_S^*(t)$ with respect to $\pi^0(t)$ is positive if

$$\gamma m_I(t)\left(V_I^*(t+1) - V_S^*(t+1)\right) > 1, \qquad \text{(COND-BR=0)}$$

in which case the best response at time t is zero, whereas the derivative of $V_S^*(t)$ with respect to $\pi^0(t)$ is negative when

$$\gamma m_I(t)\left(V_I^*(t+1) - V_S^*(t+1)\right) < 1.$$

In case of an equality, the best response can be any value between zero and one. In that case we will consider that Player 0 will decide not to confine, and therefore we have that the best response at time t is one if and only if

$$\gamma m_I(t)\left(V_I^*(t+1) - V_S^*(t+1)\right) \leq 1, \qquad \text{(COND-BR=1)}$$

Hence, Player 0 has to make a binary choice between not confining and confining.

Prior to focus on our analysis of the formulated mean field game, let us present the following result that characterizes the value of $V_I^*(t)$ since it will be useful in the analysis of our work.

Lemma 1. *We have that for* $t = 0, 1, \ldots, T - 1$,

$$V_I^*(t) = c_I \sum_{i=0}^{T-1-t} (1-\rho)^i = c_I \frac{1 - (1-\rho)^{T-t}}{\rho}.$$

Proof. See Appendix A. □

Unfortunately, we could not provide a closed-form expression for $V_S^*(t)$ for any t because $V_S^*(t)$ depends on the best-response strategy at every time larger than t. This makes this model extremely difficult to analyze. However, in the following section, we manage to provide sufficient conditions for the existence of a constant mean field equilibrium and, in the next one, of a mean field equilibrium with one jump.

4 Existence of a Constant Mean Field Equilibrium

We aim to study the conditions under which there exists a mean field equilibrium that is always one, i.e., a mean field equilibrium π such that $\pi(t) = 1$ for every $t = 0, \ldots, T$. We first show that, if $c_I \leq c_L - 1$, then $V_I^*(t+1)$ is less or equal than $V_S^*(t+1)$ for all t.

Lemma 2. *When $c_I \leq c_L - 1$, we have that $V_S^*(t) \geq V_I^*(t)$ for all $t = 0, 1, \ldots, T$.*

Proof. See Appendix B. □

From this lemma, we have that, when $c_I \leq c_L - 1$, $\gamma m_I(t)$ $(V_I^*(t+1) - V_S^*(t+1))$ is always non-positive and, as a consequence, the condition (COND-BR=1) is satisfied for all t. Therefore, we have the following result.

Proposition 1. *When $c_I \leq c_L - 1$, the best response to any π is constant.*

We now focus on the difference between $V_S^*(t)$ and $V_I^*(t)$ and we provide an upper bound of this difference.

Lemma 3. *Assume that the best response to any π at time step t is one. Therefore,*

$$V_S^*(t) - V_I^*(t) < c_I \left(1 + \frac{1 - (1-\rho)^{T-1}}{\rho}\right) - c_L + 1.$$

Proof. See Appendix C □

From the above result, we now establish a sufficient condition for the best response to any π to be constant.

Proposition 2. *Let $c_I > c_L - 1$. When*

$$\gamma m_I(0) \left(c_I \left(1 + \frac{1 - (1-\rho)^{T-1}}{\rho}\right) - c_L + 1\right) \leq 1,$$

the best response to any π is constant.

Proof. See Appendix D. □

We now note that

$$\gamma m_I(0) \left(c_I \left(1 + \frac{1 - (1-\rho)^{T-1}}{\rho}\right) - c_L + 1\right) \leq 1 \iff$$

$$c_I \leq \frac{\rho(1 + \gamma m_I(0)(c_L - 1))}{\gamma m_I(0)\left(1 + \rho - (1-\rho)^{T-1}\right)}.$$

Therefore, according to the above result, the best response to any π is constant when the cost c_I is larger than $c_L - 1$ and less or equal to $\frac{\rho(1+\gamma m_I(0)(c_L-1))}{\gamma m_I(0)(1+\rho-(1-\rho)^{T-1})}$ or, according to Proposition 1, when $c_I \leq c_L - 1$. This means that, under any of these conditions, if we consider that π is constant, the best response to π is constant (according to what we have discussed above), i.e., a constant strategy is a fixed point for the best response function. Thus, according to Definition 1, it follows that a constant strategy is a mean field equilibrium.

Let

$$c_I \leq \max\left(c_L - 1, \frac{\rho(1+\gamma m_I(0)(c_L-1))}{\gamma m_I(0)\,(1+\rho-(1-\rho)^{T-1})}\right) \qquad \text{(COND-CONST)}$$

We now present the main result of this section, which provides conditions under which a constant mean field equilibrium exists.

Proposition 3. *There exists a mean field equilibrium that is constant when* (COND-CONST) *holds.*

According to this result, we conclude that a mean field equilibrium is constant when the cost of infection is small. This means that no rational player, in this case, will get benefit of changing unilaterally the confinement strategy at any time.

In the next section, we focus on a mean field equilibrium that has one jump. We will thus assume that (COND-CONST) is not satisfied.

5 Existence of a Mean Field Equilibrium with One Jump

We first analyze the conditions under which the best response has one jump, i.e., there exists $t_0 < T-1$ such that the best response is

$$\begin{cases} 1 & \text{if } t > t_0 \\ 0 & \text{if } t \leq t_0. \end{cases}$$

We say that a strategy has, at most, one jump when it has one jump or it is constant. Let us now present the following condition that will be required to ensure that the best response has, at most, one jump.

$$c_I \geq \frac{c_L}{(1-\rho)^{T-1}}. \qquad \text{(COND1-JUMP)}$$

We now show the following result.

Proposition 4. *If* (COND1-JUMP) *holds, then the best response to any π has, at most, one jump.*

Proof. See Appendix E. □

We now focus on the existence of a mean field equilibrium, i.e., we aim to show that there exists a strategy that is a fixed point for the best response function. We consider that $\overline{\pi}$ is the strategy that is a vector with all zeros, i.e., $\overline{\pi}(t) = 0$ for all $t = 0, \ldots, T$. Let $\widetilde{\pi} \in BR(\overline{\pi})$. When (COND1-JUMP) holds, we know that $\widetilde{\pi}$ has, at most, one jump. In the remainder of this section, when we consider the strategy $\widetilde{\pi}$, we denote by $\widetilde{V}_S^*(t)$ the cost of being susceptible and $\widetilde{m}_S$ and $\widetilde{m}_I$ the proportion of the susceptible and infected population. As we showed in Lemma 1, $V_I^*(t)$ the cost of being infected does not depend on the population's strategy, and therefore when we consider the strategy $\widetilde{\pi}$ the cost of being infected does not change, and we can still denote it as $V_I^*(t)$. We aim to provide conditions such that $\widetilde{\pi} \in BR(\widetilde{\pi})$.

Let t_0 be such that $\widetilde{\pi}(t) = 1$ for all $t > t_0$ and $\widetilde{\pi}(t) = 0$ for all $t \leq t_0$. Since we know that the best response at time $T-1$ is one always, we conclude that t_0 cannot be larger or equal to $T-1$. We assume that $t_0 \geq 0$ (in Remark 4 we deal with the case where this does not occur).

We now show the following result that will be useful to prove the existence of a mean field equilibrium with one jump.

Lemma 4. *Let* (COND1-JUMP).

- *For all* $t \geq 0$, $\widetilde{m}_I(t) \geq m_I(t)$.
- *When* $t > t_0$, $V_I^*(t) \geq V_S^*(t)$
- *When* $t > t_0$, $\widetilde{V}_S^*(t) \geq V_S^*(t)$.

Proof. See Appendix F. □

Using the above results, in the following lemma, we show that the best response to $\widetilde{\pi}$ at time $t_0 + 1$ is one.

Lemma 5. *Let* (COND1-JUMP). *The best response to* $\widetilde{\pi}$ *at time* $t_0 + 1$ *is one.*

Proof. We know that the best response to $\bar{\pi}$ is one at time $t_0 + 1$. This implies that

$$\gamma m_I(t_0+1)(V_I^*(t_0+2) - V_S^*(t_0+2)) \leq 1. \tag{3}$$

We now remark that, when $t \leq t_0$, $\bar{\pi}(t) = \widetilde{\pi}(t)$ and, as a result, $m_I(t_0+1) = \widetilde{m}_I(t_0+1)$.

Besides, using Lemma 4, we conclude that $V_I^*(t_0+2) - \widetilde{V}_S^*(t_0+2)$ is smaller or equal than $V_I^*(t_0+2) - V_S^*(t_0+2)$. Thus, it follows from (3) that

$$\gamma \widetilde{m}_I(t_0+1)(V_I^*(t_0+2) - \widetilde{V}_S^*(t_0+2)) \leq \gamma m_I(t_0+1)(V_I^*(t_0+2) - V_S^*(t_0+2)) \leq 1,$$

which according to (COND-BR=1) means that the best response to $\widetilde{\pi}$ at time $t_0 + 1$ is one. □

As a consequence of the above reasoning, we have that, when (COND1-JUMP) holds, $\widetilde{\pi}$ is a mean field equilibrium if and only if the best response to $\widetilde{\pi}$ at time t_0 is equal to zero. According to (COND-BR=0), this occurs when

$$\gamma \widetilde{m}_I(t_0)(V_I^*(t_0+1) - \widetilde{V}_S^*(t_0+1)) > 1.$$

We now notice that for $t \leq t_0$, we have that $\bar{\pi}(t) = \tilde{\pi}(t)$ and, as a result, we also derive that $m_I(t_0) = \widetilde{m}_I(t_0)$. This implies that the above expression can be alternatively written as follows:

$$\gamma m_I(t_0)(V_I^*(t_0+1) - \widetilde{V}_S^*(t_0+1)) > 1.$$

As a result,

$$\tilde{\pi} \in BR(\tilde{\pi}) \iff \gamma m_I(t_0)(V_I^*(t_0+1) - \widetilde{V}_S^*(t_0+1)) > 1.$$

We now aim to investigate the conditions under which the above expression is satisfied. Let us now present the following auxiliary result.

Lemma 6. *Let* (COND1-JUMP). *For* $t > t_0$, $V_I^*(t) - \widetilde{V}_S^*(t)$ *is decreasing with* t.

Proof. See Appendix G. □

Taking into account that the best response to any π at time $T-1$ is one and the costs at time T are zero, it follows that $V_I^*(T-1) = c_I$ and $\widetilde{V}_S^*(T-1) = c_L - 1$. Using the result of Lemma 6, we obtain the following result:

Lemma 7. *Let* (COND1-JUMP). *For all* $t > t_0$,

$$V_I^*(t) - \widetilde{V}_S^*(t) \geq V_I^*(T-1) - \widetilde{V}_S^*(T-1) = c_I - c_L + 1.$$

From this result, we conclude that the condition $\gamma m_I(t_0)(V_I^*(t_0+1) - \widetilde{V}_S^*(t_0+1)) > 1$ is satisfied when

$$\gamma m_I(t_0)(c_I - c_L + 1) > 1. \tag{4}$$

We now remark that, from Assumption 1, $m_I(t_0) > m_I(T)$. Moreover, when (COND1-JUMP) we have that $c_I > c_L - 1$ and, as a result,

$$\gamma m_I(t_0)(c_I - c_L + 1) > \gamma m_I(T)(c_I - c_L + 1).$$

Using (1), we have that $m_I(T) = (1-\rho)^T m_I(0)$. Therefore, (4) is satisfied when

$$\gamma m_I(0)(1-\rho)^T(c_I - c_L + 1) > 1. \tag{COND2-JUMP}$$

From the above reasoning, the next result follows.

Proposition 5. *A mean field equilibrium with one jump exists when* (COND1-JUMP) *and* (COND2-JUMP) *hold.*

In Sect. 6, we discuss our numerical experiments that show how (COND1-JUMP) and (COND2-JUMP) influence on the existence of a mean field equilibrium.

Let us note that

$$\gamma m_I(0)(1-\rho)^T(c_I - c_L + 1) > 1 \iff c_I > c_L - 1 + \frac{1}{\gamma m_I(0)(1-\rho)^T}.$$

This provides the following expression for the existence of a mean field equilibrium with one jump which is analogous to that of (COND-CONST) for the existence of a mean field equilibrium that is constant:

$$c_I > \max\left(c_L - 1 + \frac{1}{\gamma m_I(0)(1-\rho)^T}, \frac{c_L}{(1-\rho)^{T-1}}\right).$$

According to the derived expression, we conclude that, when c_I is large, there exists a mean field equilibrium that consists of a strategy with one jump. This means that, when players incur a high cost of being infected, they get confined at the beginning of the epidemic and they do not get confined after a fixed threshold time.

Remark 3. We now assume that (COND1-JUMP) and (COND2-JUMP) hold and we consider that the jump is given at $T-2$. According to (COND-BR=0), this occurs when

$$\gamma m_I(T-2)(V_I^*(T-1) - V_S^*(T-1)) > 1.$$

Since $V_I^*(T-1) - V_S^*(T-1) = c_I - c_L + 1$ and $m_I(T-2) = (1-\rho)^{T-2} m_I(0)$ (which holds because $\bar{\pi}$ is a vector with all zeros), the above expression is equivant to

$$\gamma(1-\rho)^{T-2} m_I(0)(c_I - c_L + 1) > 1.$$

This expression is clearly satisfied when (COND2-JUMP) is satisfied. Therefore, we conclude that when (COND1-JUMP) and (COND2-JUMP) hold, the jump is given at time $T-2$.

Remark 4. Let us consider that it does not exist a t_0 such that the best response to $\bar{\pi}$ has one jump. Hence, the best response to $\bar{\pi}$ (which is a vector with all zeros) is constant. According to (COND-BR=1), if (COND1-JUMP) holds this occurs when

$$\gamma m_I(0)(V_I^*(1) - V_S^*(1)) < 1.$$

According the result of Lemma 5, we derive that the best response at time zero is one. As a result, the best response to $\widetilde{\pi}$ is a constant strategy if $\gamma m_I(0)(V_I^*(1) - \widetilde{V}_S^*(1)) < 1$, which implies that a mean field equilibrium that is constant exists when this condition is verified. This provides an additional sufficient condition for the existence of a constant mean field equilibrium to those presented in Sect. 4.

6 Discussion of (COND1-JUMP) and (COND2-JUMP)

In Proposition 4, we established conditions under which the best response has, at most, one jump. We now aim to analyze the best response when these conditions do not hold and we show that, for this instance, the best response might have multiple jumps.

We consider the following parameters: $T = 100$, $\gamma = 0.85$, $\rho = 0.75$, $c_I = 86$, $c_L = 2$ and $m_S(0) = 0.88$ and $m_I(0) = 0.12$. It is easy to check that these parameters do not satisfy the conditions (COND1-JUMP). In Fig. 2, we consider that π is a vector of all ones and we illustrate the best response to π for the considered parameters.

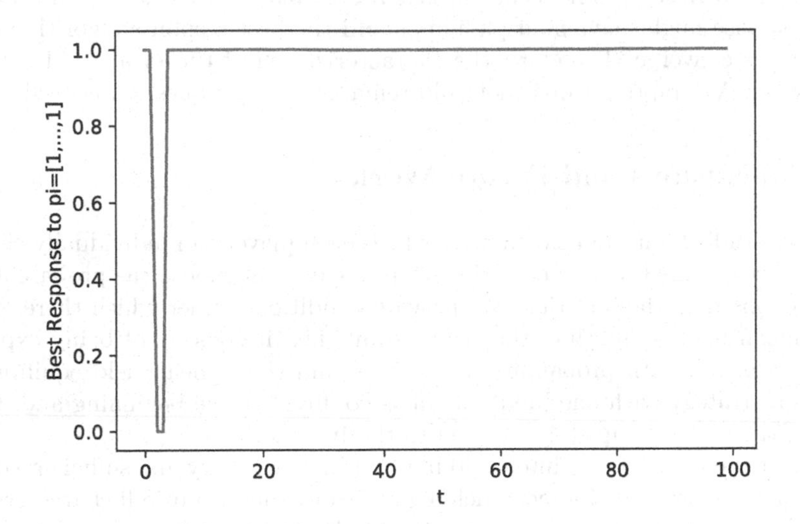

Fig. 2. The best response to $\pi = (1, \ldots, 1)$.

We observe that the best response strategy has two jumps. That is, the behavior of a selfish player consists of being exposed to the infection with probability one at the first two time steps; then, a selfish player would be confined for two time steps and, finally, the best response is equal to one until the end. This result shows that, even though the proportion of infected population decreases with t (and therefore, the maximum number of infected population is achieved at the beginning), when T is large, a selfish player might prefer to be exposed to the infection in the first two time steps instead of being confined. The main reason for this counter-intuitive behavior is that, when the epidemic is long, the player will almost surely get the infection and, therefore, it might decide to be completely exposed in the first time steps.

We also analyzed the structure of a mean field equilibrium with this set of parameters. We first computed the best response to all the strategies π with one jump as well as to the strategy π that is constant. We observed that none of these instances led to a fixed point. Therefore, we conclude that a mean field equilibrium with at most one jump does not always exist. Moreover, we also observed that the fixed point algorithm did not converge to a strategy after 300 iterations. Therefore, when (COND1-JUMP) is not satisfied, the characterization of the mean field equilibrium remains an open question.

We have observed that the fixed point algorithm converges to a mean field equilibrium with at most one jump in all the instances in which (COND1-JUMP) is satisfied, but (COND2-JUMP) not. This numerical work suggests that the mean field equilibrium is constant or has a single jump when (COND1-JUMP) is satisfied. Therefore, (COND1-JUMP) seems to be a necessary and sufficient condition for the existence of a mean field equilibrium with at most one jump.

Finally, we have also studied numerically the structure of a mean field equilibrium when Assumption 1 is not satisfied. For this case, we have seen that the best response might have multiple jumps and the best response algorithm does not always converge. Therefore, the characterization of the mean field equilibrium when Assumption 1 does not hold remains an open question as well.

7 Conclusions and Future Work

We have studied a mean field game in which each player can individually choose how to get confined, i.e., in each time step players can choose the probability of being exposed to the infection. We provide conditions under which there exists (a) a mean field equilibrium that is constant, i.e., it consists of being exposed to the infection with probability one always and (b) a mean field equilibrium that is a strategy with one jump, i.e., it is confined at the beginning and, from a given time, it is completely exposed to the infection.

For future work, we are interested in providing necessary and sufficient conditions for the existence of a mean field equilibrium and also in full characterizing it for the considered assumptions. We would also like to analyze the efficiency of the mean field equilibrium, i.e., if the cost at the mean field equilibrium is much larger than the optimal cost in the system. Finally, we would like to explore this mean field game beyond the ending of the epidemic, for an arbitrary dynamic of the proportion of infected population (i.e., not necessarily decreasing with t).

A Proof of Lemma 1

We first note that $V_I^*(T-1) = c_I + (1-\rho)V_I^*(T) = c_I$, which clearly verifies the desired condition.

We now assume that $V_I^*(t+1) = c_I \sum_{i=0}^{T-2-t}(1-\rho)^i$, for $t < T-1$ and we verify the following:

$$V_I^*(t) = c_I + (1-\rho)c_I \sum_{i=0}^{T-2-t} (1-\rho)^i = c_I + c_I \sum_{i=1}^{T-1-t} (1-\rho)^i$$
$$= c_I \sum_{i=0}^{T-1-t} (1-\rho)^i = c_I \frac{1-(1-\rho)^{T-t}}{\rho}.$$

And the desired result follows.

B Proof of Lemma 2

Let us first observe that the desired result holds for $t = T$ since $V_S^*(T) = 0$ and $V_I^*(T) = 0$. We now note that, from Lemma 1, we have that $V_I^*(T-1) = c_I$. Besides, using that when $t = T-1$ the best response to π is always one, it follows that $V_S^*(T-1) = c_L - 1$. As a consequence, $V_S^*(T-1) \geq V_I^*(T-1)$ when $c_I \leq c_L - 1$, i.e., the desired result is also satisfied when $t = T-1$. We now assume that $V_S^*(t+1) \geq V_I^*(t+1)$ for $t < T-1$ and we verify that $V_S^*(t) \geq V_I^*(t)$ when $c_I \leq c_L - 1$.

$$\begin{aligned} V_S^*(t) &= \min_{\pi^0(t)\in[0,1]} \left[c_L - \pi^0(t) + (1-\gamma m_I(t)\pi^0(t))V_S^*(t+1) + \gamma m_I(t)\pi^0(t)V_I^*(t+1)\right] \\ &= \min_{\pi^0(t)\in[0,1]} \left[c_L - \pi^0(t) + V_S^*(t+1) + \gamma m_I(t)\pi^0(t)\left(V_I^*(t+1) - V_S^*(t+1)\right)\right] \\ &= c_L - 1 + V_S^*(t+1) + \gamma m_I(t)\left(V_I^*(t+1) - V_S^*(t+1)\right), \end{aligned}$$

where the last equality holds since $V_I^*(t+1) - V_S^*(t+1) \leq 0$ and, therefore, the value that minimizes $c_L - \pi^0(t) + (1-\gamma m_I(t)\pi^0(t))V_S^*(t+1) + \gamma m_I(t)\pi^0(t)V_I^*(t+1)$ is $\pi^0(t) = 1$. Since $V_I^*(t) = c_I + (1-\rho)V_I^*(t+1) < c_I + V_I^*(t+1)$, it follows that

$$\begin{aligned} V_I^*(t) - V_S^*(t) &< c_I + V_I^*(t+1) - \left(c_L - 1 + V_S^*(t+1) + \gamma m_I(t)\left(V_I^*(t+1) - V_S^*(t+1)\right)\right) \\ &= c_I - c_L + 1 + (1-\gamma m_I(t))\left(V_I^*(t+1) - V_S^*(t+1)\right), \end{aligned}$$

which is clearly non-positive because $c_I \leq c_L - 1$ and $\gamma m_I(t) < 1$ and since we assumed that $V_S^*(t+1) \geq V_I^*(t+1)$. And the desired result follows.

C Proof of Lemma 3

Since the best response at time t is one,

$$V_S^*(t) = c_L - 1 + (1-\gamma m_I(t))V_S^*(t+1) + \gamma m_I(t)V_I^*(t+1).$$

Besides,

$$V_I^*(t) = c_I + (1-\rho)V_I^*(t+1) < c_I + V_I^*(t+1).$$

As a result,

$$\begin{aligned} V_I^*(t) - V_S^*(t) &< c_I + V_I^*(t+1) - c_L + 1 - (1-\gamma m_I(t))V_S^*(t+1) - \gamma m_I(t)V_I^*(t+1) \\ &< c_I + V_I^*(t+1) - c_L + 1 - \gamma m_I(t)V_I^*(t+1) \\ &< c_I + V_I^*(t+1) - c_L + 1 \\ &= c_I\left(1 + \frac{1-(1-\rho)^{T-t-1}}{\rho}\right) - c_L + 1 \\ &< c_I\left(1 + \frac{1-(1-\rho)^{T-1}}{\rho}\right) - c_L + 1. \end{aligned}$$

And the desired result follows.

D Proof of Proposition 2

We know that the best response to π is one when $t = T-1$ because the costs at time T are zero. Therefore, we only need to show that the best response to any π is one for all $t < T-1$.

We assume that the best response to any π is one at time $t+1$. Therefore, from Lemma 3, it follows that

$$V_S^*(t+1) - V_I^*(t+1) < c_I\left(1 + \frac{1-(1-\rho)^{T-1}}{\rho}\right) - c_L + 1.$$

As a result,

$$\gamma m_I(t)\left(V_I^*(t+1) - V_S^*(t+1)\right) < \gamma m_I(t)\left(c_I\left(1 + \frac{1-(1-\rho)^{T-1}}{\rho}\right) - c_L + 1\right).$$

From Assumption 1, it follows that $m_I(t) < m_I(0)$ and therefore, the rhs of the above expression is upper bounded by

$$\gamma m_I(0)\left(c_I\left(1 + \frac{1-(1-\rho)^{T-1}}{\rho}\right) - c_L + 1\right)$$

because $c_I\left(1 + \frac{1-(1-\rho)^{T-1}}{\rho}\right) - c_L + 1 > 0$ since $c_I > c_L - 1$ and $\rho > 0$. Since we have that

$$\gamma m_I(0)\left(c_I\left(1 + \frac{1-(1-\rho)^{T-1}}{\rho}\right) - c_L + 1\right) \leq 1,$$

from the above reasoning, it follows that

$$\gamma m_I(t)\left(V_I^*(t+1) - V_S^*(t+1)\right) < 1,$$

which, according to (COND-BR=1), it implies that the best response to any π at time t is one. And the desired result follows.

E Proof of Proposition 4

Let us recall that the best response to any π at time $T-1$ is one. We aim to show that, when (COND1-JUMP) holds, the best response to π does not have more than one jump. For a fixed strategy π, let t_0 be the first time (starting from T) such that the best response to π is zero. The desired result follows if we show that for all $t \leq t_0$ the best response to π is zero. Using an induction argument, we assume that, there exists a $\bar{t} \leq t_0$ such that the best response to π at time $\bar{t}$ is zero, which according to (COND-BR=0) is achieved when

$$\gamma m_I(\bar{t})(V_I^*(\bar{t}+1) - V_S^*(\bar{t}+1)) > 1 \tag{5}$$

and we aim to show that

$$\gamma m_I(\bar{t}-1)(V_I^*(\bar{t}) - V_S^*(\bar{t})) > 1 \tag{6}$$

i.e., that the best response to π at time $\bar{t}-1$ is zero as well. Since the best response to π at time $\bar{t}$ is zero, it follows that

$$V_S^*(\bar{t}) = c_L + V_S^*(\bar{t}+1),$$

and we also have that $V_I^*(\bar{t}) = c_I(1-\rho)^{T-\bar{t}-1} + V_I^*(\bar{t}+1)$. As a result,

$$V_I^*(\bar{t}) - V_S^*(\bar{t}) = c_I(1-\rho)^{T-\bar{t}-1} - c_L + V_I^*(\bar{t}+1) - V_S^*(\bar{t}+1).$$

From (5), we obtain that $V_I^*(\bar{t}+1) - V_S^*(\bar{t}+1) > \frac{1}{\gamma m_I(\bar{t})}$, therefore

$$V_I^*(\bar{t}) - V_S^*(\bar{t}) > c_I(1-\rho)^{T-\bar{t}-1} - c_L + \frac{1}{\gamma m_I(\bar{t})}$$

From (COND1-JUMP), we derive that $c_I(1-\rho)^{T-\bar{t}-1} - c_L > 0$, which means that the rhs of the above expression is lower bounded by

$$V_I^*(\bar{t}) - V_S^*(\bar{t}) > \frac{1}{m_I(\bar{t})}.$$

We multiply both sides by $\gamma m_I(\bar{t}-1)$:

$$\gamma m_I(\bar{t}-1)(V_I^*(\bar{t}) - V_S^*(\bar{t})) > \frac{m_I(\bar{t}-1)}{m_I(\bar{t})}.$$

We now notice that $\frac{m_I(\bar{t}-1)}{m_I(\bar{t})} > 1$ due to Assumption 1 and, therefore, (6) holds which implies that the desired result follows.

F Proof of Lemma 4

F.1 $\widetilde{m}_I(t) \geq m_I(t)$

We first show that $\widetilde{m}_I(t) \geq m_I(t)$ for $t \geq 0$. We note that, at time zero, both values coincide, i.e., $\widetilde{m}_I(0) = m_I(0)$. We now assume that for $t \geq 0$, $\widetilde{m}_I(t) \geq m_I(t)$, and we aim to show that $\widetilde{m}_I(t+1) \geq m_I(t+1)$. Using (1) and also that $\bar{\pi}(t)$ is formed by all zeros,

$$m_I(t+1) = m_I(t)(1 + \gamma m_S(t)\bar{\pi}(t) - \rho) = m_I(t)(1-\rho),$$

whereas for $\widetilde{\pi}$ we have

$$\widetilde{m}_I(t+1) = \widetilde{m}_I(t)(1+\gamma\widetilde{m}_S(t)\widetilde{\pi}(t)-\rho) \geq \widetilde{m}_I(t)(1-\rho) \geq m_I(t)(1-\rho) = m_I(t+1).$$

And the desired result follows.

F.2 $V_I^*(t) \geq V_S^*(t)$

We now focus on the proof of $V_I^*(t) \geq V_S^*(t)$ for all $t > t_0$. We first note that $V_I^*(T) = V_S^*(T) = 0$ and, therefore, the desired result follows at time T. We now assume that $V_I^*(t+1) \geq V_S^*(t+1)$ for $t+1 > t_0$ and we aim to show that $V_I^*(t) \geq V_S^*(t)$. Since the best response to $\bar{\pi}$ at time t is one, we have that

$$V_S^*(t) = c_L - 1 + (1 - \gamma m_I(t))V_S^*(t+1) + \gamma m_I(t)V_I^*(t+1)$$

and taking into account that $V_I^*(t) = c_I(1-\rho)^{T-1-t} + V_I^*(t+1)$, we have that

$$V_I^*(t) - V_S^*(t) = c_I(1-\rho)^{T-1-t} - c_L + 1 + (1-\gamma m_I(t))(V_I^*(t+1) - V_S^*(t+1)) \geq 0,$$

which holds since $V_I^*(t+1) \geq V_S^*(t+1)$ and $c_I \geq \frac{c_L}{(1-\rho)^{T-1}} > \frac{c_L}{(1-\rho)^{T-t-1}}$. And the desired result follows.

F.3 $\widetilde{V}_S^*(t) \geq V_S^*(t)$

Finally, we show that $\widetilde{V}_S^*(t) \geq V_S^*(t)$ for all $t > t_0$. We know that $\widetilde{V}_S^*(T) = V_S^*(T) = 0$ since the cost at the end is zero. Therefore, we assume that $\widetilde{V}_S^*(t+1) \geq V_S^*(t+1)$ for $t > t_0$ and we aim to show that $\widetilde{V}_S^*(t) \geq V_S^*(t)$.

We know that the best response to $\bar{\pi}$ for $t > t_0$ is one. Therefore, $V_S^*(t) = c_L - 1 + V_S^*(t+1) + \gamma m_I(t)(V_I^*(t+1) - V_S^*(t+1))$. For $\widetilde{V}_S^*(t)$, we denote by a the best response to $\widetilde{\pi}$ at time t. Thus,

$$\widetilde{V}_S^*(t) = c_L - a + (1 - \gamma\widetilde{m}_I(t)a)\widetilde{V}_S^*(t+1) + \gamma\widetilde{m}_I(t)aV_I^*(t+1).$$

Using that $\widetilde{V}_S^*(t+1) \geq V_S^*(t+1)$ and because $1 - \gamma\widetilde{m}_I(t)a$ is positive, we get

$$\widetilde{V}_S^*(t) \geq c_L - a + (1 - \gamma\widetilde{m}_I(t)a)V_S^*(t+1) + \gamma\widetilde{m}_I(t)aV_I^*(t+1). \tag{7}$$

Therefore, $\widetilde{V}_S^*(t) \geq V_S^*(t)$ is true when

$$\begin{aligned} &c_L - a + (1 - \gamma\widetilde{m}_I(t)a)V_S^*(t+1) + \gamma\widetilde{m}_I(t)aV_I^*(t+1) \\ &\geq c_L - 1 + V_S^*(t+1) + \gamma m_I(t)(V_I^*(t+1) - V_S^*(t+1)). \end{aligned}$$

Simplifying

$$1 - a \geq \gamma(m_I(t) - \widetilde{m}_I(t)a)(V_I^*(t+1) - V_S^*(t+1)).$$

Using that $\widetilde{m}_I(t) \geq m_I(t)$ and since $V_I^*(t+1) - V_S^*(t+1)$ is non-negative, the rhs of the above expression is smaller or equal than $\gamma(1-a)m_I(t)(V_I^*(t+1) - V_S^*(t+1))$. Therefore, a sufficient condition for the desired result to hold is

$$1 - a \geq \gamma(1-a)m_I(t)(V_I^*(t+1) - V_S^*(t+1)).$$

We now differentiate two cases: (a) when $a = 1$, we have zero in both sides of the expression and therefore, the condition is satisfied; (b) when $a \neq 1$, we divide by $1 - a$ both sides of the expression and we get

$$1 \geq \gamma m_I(t)(V_I^*(t+1) - V_S^*(t+1)),$$

which is also satisfied from (COND-BR=1) because the best response to $\bar{\pi}$ at time t is one.

G Proof of Lemma 6

We aim to show that

$$V_I^*(t) - \widetilde{V}_S^*(t) \geq V_I^*(t+1) - \widetilde{V}_S^*(t+1),$$

for $t > t_0$. From Lemma 5 we know that the best response to $\widetilde{\pi}$ is one for $t > t_0$, because (COND1-JUMP) is satisfied, and therefore,

$$\widetilde{V}_S^*(t) = c_L - 1 + \widetilde{V}_S^*(t+1) + \gamma m_I(t)(V_I^*(t+1) - \widetilde{V}_S^*(t+1)).$$

From Lemma 1, we have that

$$V_I^*(t) = c_I(1-\rho)^{T-t-1} + V_I^*(t+1).$$

Therefore,

$$V_I^*(t) - \widetilde{V}_S^*(t) = c_I(1-\rho)^{T-t-1} - c_L + 1 + (1 - \gamma m_I(t))(V_I^*(t+1) - \widetilde{V}_S^*(t+1)).$$

Hence, using the above expression, we get that

$$\begin{aligned} V_I^*(t) - \widetilde{V}_S^*(t) \geq V_I^*(t+1) - \widetilde{V}_S^*(t+1) \iff \\ c_I(1-\rho)^{T-t-1} - c_L + 1 - \gamma m_I(t)(V_I^*(t+1) - \widetilde{V}_S^*(t+1)) \geq 0. \end{aligned}$$

We now note that when $c_I \geq \frac{c_L}{(1-\rho)^{T-1}}$, we get $c_I(1-\rho)^{T-t-1} > c_L$, and as the best response to $\widetilde{\pi}$ is one we have (COND-BR=1). Therefore,

$$\begin{aligned} c_I(1-\rho)^{T-t-1} - c_L + 1 - \gamma m_I(t)(V_I^*(t+1) - \widetilde{V}_S^*(t+1)) \\ > 1 - \gamma m_I(t)(V_I^*(t+1) - \widetilde{V}_S^*(t+1)) \geq 0 \end{aligned}$$

and the desired result follows.

References

1. Anderson, R.M., May, R.M.: Infectious Diseases of Humans: Dynamics and Control. Oxford University Press, Oxford (1992)
2. Cabannes, T., et al.: Solving n-player dynamic routing games with congestion: a mean field approach. arXiv preprint arXiv:2110.11943 (2021)
3. Cho, S.: Mean-field game analysis of sir model with social distancing. arXiv preprint arXiv:2005.06758 (2020)
4. Daskalakis, C., Goldberg, P.W., Papadimitriou, C.H.: The complexity of computing a Nash equilibrium. SIAM J. Comput. **39**(1), 195–259 (2009)
5. Diekmann, O., Andre, J., Heesterbeek, P.: Mathematical Epidemiology of Infectious Diseases: Model Building, Analysis and Interpretation, vol. 5. Wiley, Hoboken (2000)
6. Doncel, J., Gast, N., Gaujal, B.: Discrete mean field games: existence of equilibria and convergence. J. Dyn. Games **6**(3), 1–19 (2019)

7. Doncel, J., Gast, N., Gaujal, B.: A mean field game analysis of sir dynamics with vaccination. Probab. Eng. Inf. Sci. **36**(2), 482–499 (2022)
8. Elie, R., Mastrolia, T., Possamaï, D.: A tale of a principal and many, many agents. Math. Oper. Res. **44**(2), 440–467 (2019)
9. Ghilli, D., Ricci, C., Zanco, G.: A mean field game model for Covid-19 with human capital accumulation. arXiv preprint arXiv:2206.04004 (2022)
10. Huang, K., Di, X., Du, Q., Chen, X.: A game-theoretic framework for autonomous vehicles velocity control: bridging microscopic differential games and macroscopic mean field games. arXiv preprint arXiv:1903.06053 (2019)
11. Huang, M., Malhamé, R.P., Caines, P.E.: Large population stochastic dynamic games: closed-loop Mckean-Vlasov systems and the Nash certainty equivalence principle. Commun. Inf. Syst. **6**(3), 221–252 (2006)
12. Hubert, E., Turinici, G.: Nash-MFG equilibrium in a sir model with time dependent newborn vaccination (2016)
13. Kermack, W.O., McKendrick, A.G.: A contribution to the mathematical theory of epidemics. Proc. Roy. Soc. Lond. Ser. A Containing Papers of a Mathematical and Physical Character, **115**(772), 700–721 (1927)
14. Kolokoltsov, V.N., Malafeyev, O.A.: Corruption and botnet defense: a mean field game approach. Int. J. Game Theory **47**(3), 977–999 (2018)
15. Laguzet, L., Turinici, G.: Global optimal vaccination in the sir model: properties of the value function and application to cost-effectiveness analysis. Math. Biosci. **263**, 180–197 (2015)
16. Lasry, J.-M., Lions, P.-L.: Jeux à champ moyen. i-le cas stationnaire. Comptes Rendus Mathématique **343**(9), 619–625 (2006)
17. Lasry, J.-M., Lions. , P.-L.: à champ moyen. ii-horizon fini et contrôle optimal. Comptes Rendus Mathématique **343**(10), 679–684 (2006)
18. Lasry, J.-M., Lions, P.-L.: Mean field games. Japan. J. Math. **2**(1), 229–260 (2007)
19. Olmez, S.Y., et al.: How does a rational agent act in an epidemic? arXiv e-prints, pp. arXiv-2206 (2022)

Data Center Organization and Optimization Strategy as a K-Ary N-Cube Topology

Pedro Juan Roig[1](✉), Salvador Alcaraz[1], Katja Gilly[1], Cristina Bernad[1], and Carlos Juiz[2]

[1] Miguel Hernández University (Elche), Elche, Spain
{proig,salcaraz,katya,cbernad}@umh.es
[2] University of the Balearic Islands (Palma de Mallorca), Palma, Spain
cjuiz@uib.es

Abstract. Data center deployments are increasing because of the ongoing expansion of IoT environments, which require data centers of a small to medium size, where forwarding traffic is optimized in order to save energy consumption. In this paper, a data center organization and optimization strategy by using a k-ary n-cube topology has been proposed as such a design provides an easy manner to deal with traffic, which also has the advantage of presenting a collection of redundant paths to undertake alternative ways between any pair of nodes, thus avoiding single points of failure. Furthermore, a formal algebraic model referred to this topology has also been specified and verified.

Keywords: Data center design · Formal algebraic model · k-ary n-cube · Resource migration · Toroidal topology

1 Introduction

Edge computing deployments are growing sharply due to the expansion of Internet of Things (IoT) ecosystems [1], with Artificial Intelligence (AI) playing a crucial part in improving performance [2]. Such IoT ecosystems may be applied to basically any sector, such as urban structure, energy generation or resource extraction, even though health care and manufacturing get the biggest stake [3].

The implementation of IoT domain will lead to economic growth, as it appears to be a significant correlation between IoT deployments and Total Factor Productivity (TFP) [4], in spite of IoT being in the early stages of its development, although some projections point out huge rates of growth up to 2025 regarding the number of IoT devices, which may reach up to 75 billion [6], as well as concerning the financial impact on the global economy, which may get up to $11.1 trillion [5]. Hence, it seems clear that activities related to IoT field may cause a great impact in Gross Domestic Product (GDP) in the coming years [7].

M. Forshaw et al. (Eds.): PASM 2022, CCIS 1786, pp. 81–92, 2023.
https://doi.org/10.1007/978-3-031-44053-3_5

On the other hand, sustainability plays an important part in IoT deployments as those may be geared to reduce carbon emissions [8], as electrical power consumption in data centers highy contributes to enlarge the carbon footprint [9]. Hence, a convenient strategy is to employ renewable energies as opposed to fossil fuel-based power [10], which may lead to a more sustainable development.

Hence, there is a clear tendency to move towards green IoT (G-IoT), being seen as a kind of an environmentally friendly IoT [11], where efficiency in energy production and consumption are key players in order to achieve sustainable digital transformation and environmental protection [12], thus unleashing the potential of IoT as a convenient technology to fulfill the 17 targets included in the Sustainable Development Goals (SDG) sponsored by UN 2030 Agenda [13].

It is to be noted that IoT may play an important part within circular economy (CE), hence helping tackle climate change, biodiversity loss, waste and pollution [14], leading to a resilient system being good for business, people and environment [15], where recycling and waste valorization may drive to crack down on waste and pollution, circulate products or materials, and regenerate nature [16].

In fact, the advent of IoT-based business models adapted to CE significantly enhances both environmental and economic performance [17], as well as improving long-term sustainability goals [18], because IoT acts as a driver for CE due to its new ways of information sharing, creating a significant and positive effect on green manufacturing, recycling or remanufacturing, and circular design [19].

In this context, data center organization and optimization is crucial in order to optimize both performance and sustainability rates [20], where a tradeoff may often need to be made between simplicity and performance. Several designs are possible in a data center, although in this paper, the focus is going to be set in a k-ary n-cube topology, which might be compared to a mesh-based network on chip scheme [21], that representing a scalable and fit solution for communications.

The rest of the paper is organized as follows: first, Sect. 2 introduces the features of toroidal shapes, after that, Sect. 3 presents the k-ary n-cube topology applied to Data Centers, in turn, Sect. 4 proposes a formal algebraic model of the aforementioned topology, afterwards, Sect. 5 carries out the verification of that model, and finally, Sect. 6 draws the final conclusions out of the paper.

2 Toroidal Shapes

To start with, toroidal arrays are n-dimensional arrays where elements in a fixed dimension are all sequentially linked, as well as the first one and the last one are linked together through a wraparound link [22]. Figure 1 depicts a 2D toroidal array with dimensions 4×4, where each row and each column are closed loops.

On the other hand, if a further dimension is added up, it results in a 3D toroidal array, which is also known as 3D hypertoroidal array, where dimensions are distributed into layers, rows and columns. Figure 2 exhibits a 3D toroidal array with dimensions $2 \times 2 \times 2$, where wraparound links establish closed loops.

Toroidal structures may be used as the foundations of different mathematical objects, where each of them identify the nodes within the structure in a particular

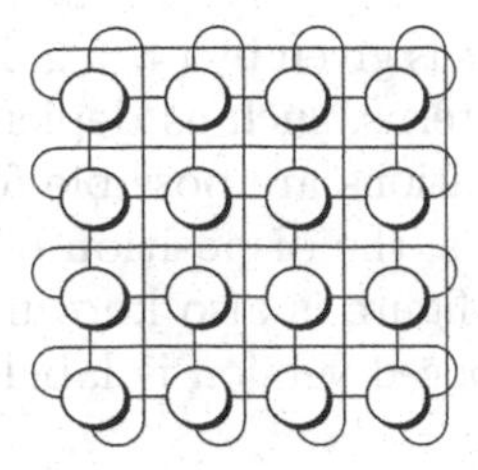

Fig. 1. Instance of a bidimensional toroidal array, with wraparound links closing each row and each column.

manner. In this sense, the most well known toroidal designs identify their nodes by means of 2 variables, where k stands for the k-ary alphabet being employed in the naming of those nodes, whilst n does for the number of dimensions available. However, each of those diverse mathematical objects will identify the nodes in a specific fashion, where k and n are combined in a given manner for each case.

One of the most well known layouts with toroidal structures are de Bruijn tori, also called de Bruijn hypertori if $n > 2$. Those are extensions of de Bruijn sequences, whose main particularity is the fact that they contain all available strings of n-length being made with a k-alphabet precisely once. In other words, a de Bruijn sequence is a k^n-length superstring, which is obviously unidimensional, having embedded all available n-length strings exactly once, in a way that all possible linear patterns with k-alphabet and n-length are present only once.

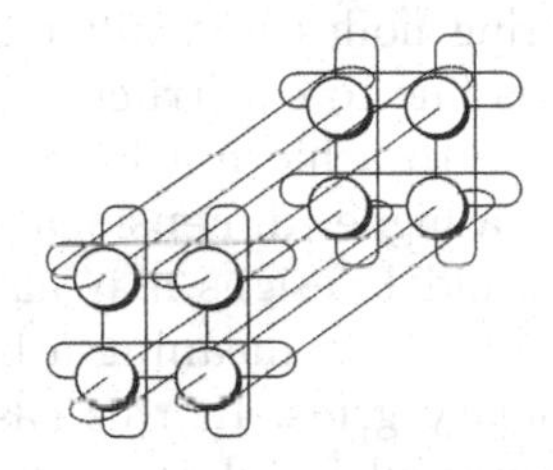

Fig. 2. Instance of a tridimensional toroidal array, with wraparound links closing each layer, each row and each column.

Hence, a de Bruijn torus represents a bidimensional shape where all available bidimensional patterns are spotted just once, the former being a toroidal matrix and the latter being its toroidal submatrices, where both the shape and the patterns may be either square or rectangular, in a way that all combinations are available. The most well known condition for a de Bruijn torus to exist is given in (1), where r_i stand for the values of dimension i of the matrix, ranged from 0 to $n-1$, whilst m_i does for the values of dimension i of the patterns, whereas k identifies the alphabet being used. Taking that all into account, a de Bruijn torus is expressed as $(r_i; m_i)_k$, where each value of i is separated by commas.

$$\prod r_i = k^{\prod m_i} \tag{1}$$

The smallest de Bruijn torus is given by $(4, 4; 2, 2)_2$, which represents a square matrix with square binary patterns, such as depicted in Fig. 3. It is to be mentioned that only two representations are possible for that object, one being the transposed of the other. Besides, the disposition of 1 s in the matrix reminds of a Brigid's cross, hence such a figure is also known as the clockwise instance of $(4, 4; 2, 2)_2$, whereas the transposed version is labeled as the counterclockwise.

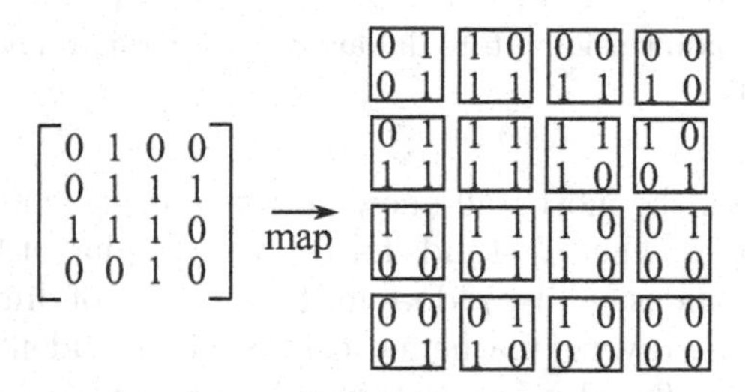

Fig. 3. Mapping of a clockwise de Bruijn torus $(4, 4; 2, 2)_2$.

It is to be noted that de Bruijn torus assign a single symbol within the k-ary alphabet to each node, where the association of nodes so as to form patterns result unique and are only repeated once throughout the whole shape.

Another example of toroidal shapes are k-ary grids, where nodes are identified with a unique string of length n, being built up with a k-alphabet, and whose main feature is that neighboring nodes just differ in a single symbol. It is to be noted that any value of k is fine to construct a k-ary grid, even though the value of n must always be even to achieve a bidimensional toroidal structure, or otherwise, if n is odd, then a higher-dimensional toroidal shape is obtained, although bidimensional non-toroidal designs may also be attained.

In the particular case where $k = 2$, meaning a binary alphabet, then those shapes may be denoted as binary grids. In this case, the node identifiers are taken from the vertices of n-hypercubes, where each nodes identifiers present as many bits as the number of dimensions n. One of the nodes being taken as a reference, which is labeled with all 0 s, and then, if another node in any position maintain the same value as the reference for a given dimension, then that bit is set to 0, or otherwise, it is set to 1. Figure 4 exposes a binary grid with $n = 4$, where specific patterns may be spotted in each row and each column. It is to be noted that the toroidal nature of the structure is easily detected.

Extending the binary case to a generic k-ary case means that the n-hypercube is transformed into a k-ary n-cube, where the values representing a given dimension of a node range from 0 to $k - 1$, whereas dimensions move from 0 to $n - 1$. Regarding k-ary grids, Fig. 5 exhibits a k-ary grid with a ternary alphabet, this is $k = 3$, and with two dimensions, such that $n = 2$, where defined patterns may be seen in each row and column and the toroidal structure is also present.

A further example of toroidal shapes are k-ary n-cube themselves, where node identifiers are the same as those cited for k-ary grids, although the layout

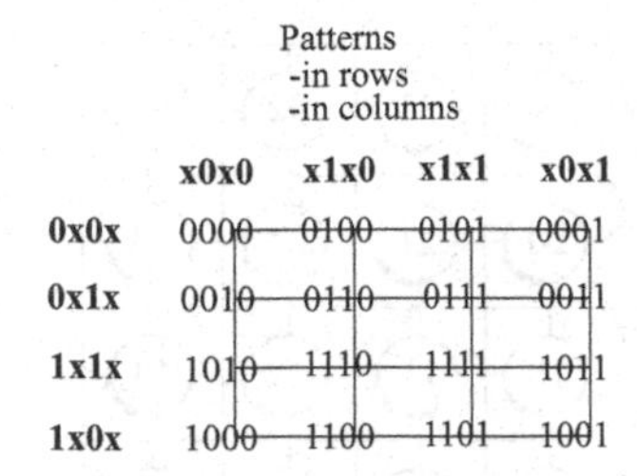

Fig. 4. Binary grid for $k = 2$ and $n = 4$.

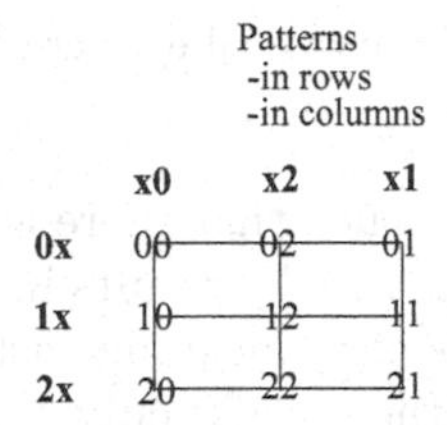

Fig. 5. k-ary grid for $k = 3$ and $n = 2$.

of those nodes differ, whilst the number of nodes is the same, that being k^n. In this case, the values for each dimension from 0 to $n-1$ grow sequentially from 0 to $k-1$, where a reference node has all its n symbols set to 0. Obviously, if the alphabet used is the binary, then all n bits are also set to 0 for reference node.

Hence, each node is identified by a string where each digit d_i denotes the value for dimension i, ranging from 0 to $n-1$, such as $d_0 \cdots d_i \cdots d_{n-1}$, where each of those values range from symbol 0 to symbol $k-1$, such as d_i is the value corresponding to dimension i. Those nodes go connected to the other nodes whose identifiers in a particular dimension i are just their predecessor or their successor, thus only differing in $d_i \pm 1 mod\ k$. This results in an overall number of neighbors of $2n$, with just two of them per each dimension i [23].

Figure 6 represents a k-ary n-cube with $k = 4$, thus having 4 nodes per dimension, whose values in a single dimension vary from 0 to 3, as well as $n = 2$, hence having 2 dimensions, which makes for a bidimensional figure, specifically a square. Each node is identified by a string of 2 symbols, where the least significant one stands for the horizontal dimension, whilst the most significant one does for the vertical dimension. Besides, it may be appreciated that values for each digit range from 0 to 3 along the same row or column, whereas the toroidal structure means that after digit 3 ($k-1$) comes digit 0 when going rightwards or upwards, whilst the other way around happens when moving leftwards or downwards [24].

Furthermore, Fig. 7 depicts a k-ary n-cube where $k = 2$, hence having binary values for dimension, and $n = 3$, thus representing each node with a string of 3 symbols, which happens to be bits because of the binary alphabet, each of those representing one of the 3 dimensions involved. It is to be said that in the binary cases, each bit b_i describes the value for dimension i, ranging from 0 to 1, such as $b_0 \cdots b_i \cdots b_{n-1}$, where each of those values may be worth 0 or 1.

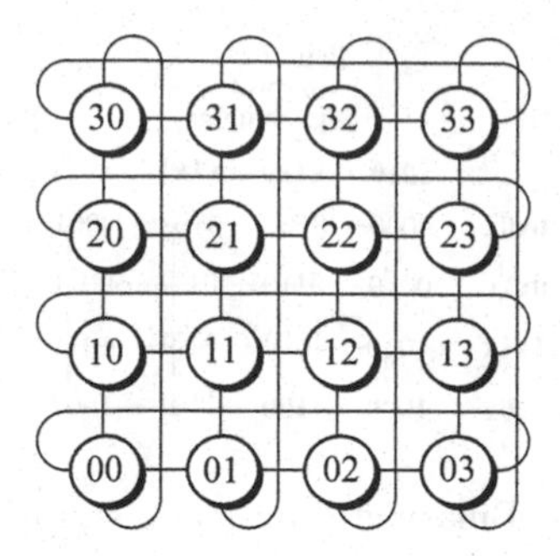

Fig. 6. Nodes of a k-ary n-cube where $k = 4$ and $n = 2$.

Additionally, it is to be mentioned that there is just one neighbor per dimension for binary alphabets, as $b_i \pm 1 mod\ 2$ results in the same value, which results in an total amount of neighbors of n, meaning just one per each dimension i.

It is to be reminded that in the k-ary n-cube structures shown in the figures, a reference node is assigned to one node in a corner, where the identifier is composed only by 0 s, and then, the first row is filled in sequentially, then the second, and so on until all the layer is complete before going to the other layers.

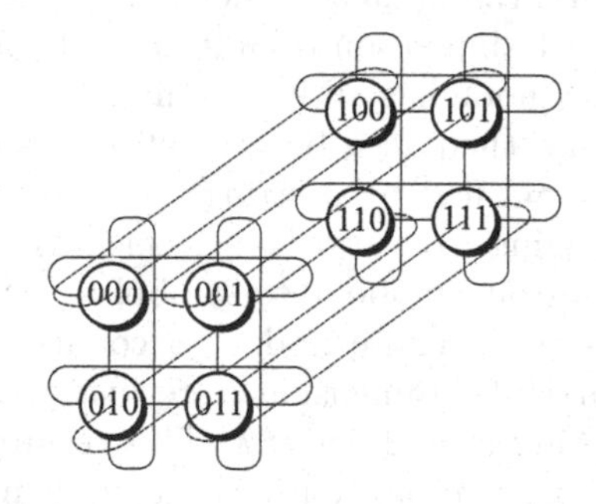

Fig. 7. Nodes of a k-ary n-cube where $k = 2$ and $n = 3$.

Moreover, it is to be noted that k-ary grids presented above and k-ary n-cubes share the same physical structure when $k \leq 3$, although their logical layouts change as node identifiers differ for both cases, whilst physical setup does not match in both cases for $k > 3$, as a single node bears $n(k-1)$ neighbors in the latter, whereas it does $2n$ in the latter, or even n in case $k = 2$.

3 A Topology Related to k-Ary n-Cube

As exposed above, IoT deployments are ever rising, which in most of the cases are associated to data center implementations for edge domains in order to optimize the processing and storage of big data being generated by IoT devices [25].

Edge data centers must generally deal with a smaller number of devices than their counterparts located in the cloud, hence their size need to be smaller, meaning less hosts have to be involved. Therefore, an interesting layout for interconnecting those hosts may be a k-ary n-cube, as several redundant paths are available to go from a source host to a destination host and adjusting the values of k and n may adapt to the different requirements of a particular scenario [26].

In order to construct a model of a data center with the layout of a k-ary n-cube for the links among physical hosts, it is to be established a model for a generic host with the number of ports available to interconnect to its neighbors. Therefore, considering a generic value of k, it is necessary to have $2n$ links, such that the port going to its predecessor host in dimension $i = \{0 \cdots k-1\}$ is $2i$, whereas the port going to its successor host in that dimension i is $2i+1$, thus ranging from port 0 to port $2n-1$, as exhibited in Fig. 8.

This way, port 0 points out to its predecessor in dimension, whilst port 1 does to its successor in such a dimension, then, port 2 goes to its predecessor, whereas port 3 does to its successor, and the same scheme keeps going by assigning two ports to each dimension until getting to dimension $k-1$, where port $2n-2$ points out to its predecessor and port $2n-1$ does to its successor.

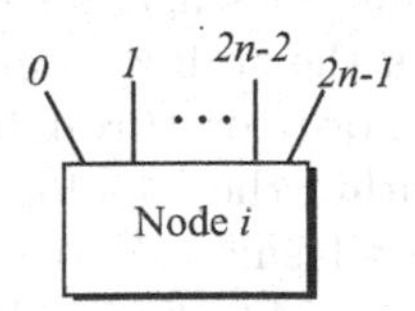

Fig. 8. k-ary n-cube topology with ports.

Additionally, it is to be noted that this type of topology may forward traffic among nodes by means of arithmetic operations, such as integer divisions and modular arithmetic, by taking advantage of the fact that all dimensions are composed by the same number of nodes. Hence, focusing on 2D, nodes located in the same row share the same result when applying integer division by k, whilst nodes located in the same column share the same outcome when applying modular arithmetic by k. Then, the way to reach certain value depends on where is the shortest distance, which is related to half k, to reach the crossing node to get there. Moreover, the same principles may be extrapolated to a third dimension regarding the layer where nodes are on, and also for higher dimensions.

4 Formal Algebraic Model of k-Ary n-Cube

In order to build up a formal algebraic model, a process algebra called Algebra of Communicating Processes (ACP) is going to be employed because of its abstract nature, which permits to take aside the real nature of objects and just focus on the relationships among them [27]. Hence, in this case, physical hosts within a k-ary n-cube are considered as objects, whilst virtual hosts are just internal pieces of the former, which may later be migrated to other physical hosts.

ACP define two atomic actions for objects, such as send a message d through channel x at port p, denoted by $s_{x,p}(d)$, as well as receive a message d through channel y at port q, described by $r_{y,q}(d)$. Then, if a send and a receive action are executed in the opposite ends of the same channel, such as $x = y$, then communication will take place in that channel, or otherwise, it those couple of actions are run in different channels, such as $x \neq y$, then those action go deadlock [28]. Furthermore, the model considers b as the destination.

Furthermore, some operators are available to describe how the atomic actions get related within an object, such as the sequential operator $(\cdot)$ to indicate that the execution of terms go sequentially, the alternate operator $(+)$ to state that diverse actions are available to choose one, the merge operator $(||)$ to point out concurrency among execution of actions, or conditional operator (*True* $\triangleleft$ *condition* $\triangleright$ *False*) to denote a condition, which may or may not be met, where the former implies the execution of one action and the latter does to run another.

Those tools are enough to model the objects of a given system, which in this case refers to a physical host within a k-ary n-cube. In this sense, expression (2) describes the model specification in ACP, where atomic actions express both the item involved and the port where a message is send to or received from [29].

Basically, the idea is to check the traffic coming into any uplink port (from $p = 0$ to $2n - 1$, as exposed above) or through an internal port (labeled as $2n$) for each of the k^n nodes within the topology (from $i = 0$ to $i = k^n - 1$), which are working in a concurrent fashion. Hence, when some traffic is detected getting into any of those ports, then the first thing to do is to inspect whether that node i is the destination one b, and if that is the case, then traffic is sent through the internal port $2n$. Otherwise, distance between i and b is evaluated in each dimension so as to see whether incoming traffic is forwarded through the precedessor link or through the successor one. Specifically, the first dimension is calculated through modular arithmetic and the rest of them are done through integer divisions by adjusting the exponents according to each dimension.

$$V_i = ||_{i=0}^{k^n-1}\Bigg(||_{p=0}^{2n}\Bigg(r_{V_i,p}(d) \cdot \Bigg(s_{V_i,2n} \triangleleft b = i \triangleright \Bigg(\Big(s_{V_i,0}(d) \triangleleft \big(\frac{-k}{2} \leq (b_{|k} - i_{|k}) < 0\big) \;\; OR \;\; \big((b_{|k} - i_{|k}) > \frac{k}{2}\big) \triangleright s_{V_i,1}(d)\Big) \cdot \Big(\sum_{t=1}^{k-1} \Big(s_{V_i,2t}(d) \triangleleft \big((\frac{-k}{2})^t \leq (\left\lfloor \frac{b}{k^t} \right\rfloor - \left\lfloor \frac{i}{k^t} \right\rfloor) < 0\big) \;\; OR \;\; \big((\left\lfloor \frac{b}{k^t} \right\rfloor - \left\lfloor \frac{i}{k^t} \right\rfloor) > (\frac{k}{2})^t\big) \triangleright s_{V_i,2t+1}(d)\Big)\Big)\Bigg)\Bigg)\Bigg)\Bigg) \cdot V_i \quad (2)$$

Regarding performance evaluation, it is to be noted that forwarding tasks in the model proposed are based on integer divisions and modular arithmetic, as opposed to searching for matches into lookup tables, which are basically undertaken by means of *AND* gates by sequentially comparing a certain table entry to each direction being stored within. On the other hand, the former needs to be implemented through specific integrated circuits aimed at calculating quotients and remainders of divisions, which are composed of a number of logic gates.

It is to be said that the hardware implementation of the relevant algorithms will determine which solution performs better. As per the model, [32] presents an interesting design consuming a relatively small amount of power and computation time for a small number of bits, which is the case of the node identifiers for the model proposed. Otherwise, [33] exposes a design to achieve an efficient lookup table obtaining lower values of computation time, although having a bit higher power consumption, as 32 bits need to be checked for each IP address (or 48 bits for each MAC address) and all of them are checked out for each entry.

5 Verification of k-Ary n-Cube Algebraic Model

Once a formal algebraic model has been designed, then it is time to apply the encapsulation operator in order to turn all internal atomic actions into communications $c_x(d)$, whilst the rest of those go deadlock δ. This operator is denoted by ∂_H, where set H contains all internal atomic actions involved in the model [30]. Furthermore, the encapsulation operator leads to show the sequence of events of a concurrent or distributed system, as just internal communications appear on the same order as they are meant to be run.

Afterwards, the abstraction operator, which is denoted by τ_I, where set I contains all internal communications involved within the model, is applied in order to mask all internal communications, thus leaving just the external actions of the model, which represents the external behavior of it, as just inputs and outputs are shown, where the model is left as a black box.

At that point, the external behavior of the real system is expressed in ACP terms, where a generic variable X is assigned to it. Then, at that point, the external behavior of the model proposed and the external behavior of the real system may be compared to each other, in a way that if both share the same string of actions and the same branching structure, then it may be said that both are rooted branching bisimilar [31], which means that there is an equivalence relation between them, which is a sufficient condition to get a model verified.

In order to apply the abstraction operator to the model, first the encapsulation operator need to be applied to the whole group of external hosts being run concurrently. As the topology selected is a k-ary n-cube, then there is an overall amount of k^n nodes, those being assigned from 0 all the way to $k^n - 1$.

It is to be observed that the model shown in (2) presents an initial receive action called $r_{V_i,p}(d)$, meaning that a message d is received into the internetworking system within the data center through a node V_i by means of port p. Then, there is a bunch of conditions to search for the shortest path to reach destination through all crossing paths available, which eventually end up on some send action towards destination called $s_{V_i,p}(d)$. Hence, the external atomic actions are just the first receive action and the send action getting to destination, whilst internal communications will get masked by the abstraction operator.

Hence, on the one hand, the external behavior of the model is shown in (3).

$$||_{i=0}^{k^n-1} \tau_I\Big(\partial_H\Big(V_i\Big)\Big) = r_{V_i,p}(d) \cdot s_{V_i,p}(d) \cdot \tau_I\Big(\partial_H\Big(V_i\Big)\Big) \tag{3}$$

On the other hand, the external behavior of the real system is seen in (4).

$$X = r_{V_i,p}(d) \cdot s_{V_i,p}(d) \cdot X \tag{4}$$

By looking at both expressions, it seems obvious that both are recursive equations where the same terms are being multiplied. Therefore, it appears to be obvious that both have the same string of actions and the same branching structure, thus an equivalence relations is established between them, which is a sufficient condition to get a model verified.

6 Conclusions

In this paper, a data center organization and optimization strategy as a k-ary n-cube topology has been proposed as an interesting solution for edge data centers, as those require a small to medium number of physical hosts, and such a layout provide a regular way to forward traffic by means of easy arithmetic operations.

To start with, an introduction about the importance of IoT, data centers and sustainability has been carried out so as to remark the importance of optimize energy consumption in data centers, which may be achieved by finding easy ways to forward traffic among hosts. Then, three toroidal shapes have been presented, where the first one has been de Bruijn torus, where all nodes are denoted by a single k-ary symbol in order to form patterns with its neighbors, the second one has been k-ary grids, where all neighboring nodes just differ in one symbol, which results in definite patterns in each value of every dimension, and the third one has been k-ary n-cubes, where nodes are sequentially assigned in rows and columns if there are just 2 dimensions, or also in layers if there is a 3rd dimension.

After that, a topology related to k-ary n-cube has been presented, and that design has been used to undertake a formal algebraic model of such a layout has been proposed, which in turn, has been verified. Performance-wise, hardware implementation will rule whether traditional lookup tables or this model prevails.

References

1. Tahaei, H., Afifi, F., Asemi, A., Zaki, F., Anuar, N.B.: The rise of traffic classification in IoT networks: a survey. J. Network Comput. Appl. **154**(C), 1–20 (2020). https://doi.org/10.1016/j.jnca.2020.102538
2. Alanezi, M.A.: An efficient framework for intelligent learning based on artificial intelligence and IoT. Int. J. Emerg. Technol. Learn. **17**(07), 112–124 (2022). https://doi.org/10.3991/ijet.v17i07.27851
3. Sivagami, P., Illavarason, P., Harikrishnan, P., Reddy, G.: IoT Ecosystem - a survey on classification of IoT. In: Proceedings of the 1st International Conference on Advanced Scientific Innovation in Science, Engineering and Technology (ICASISET 2020) 1–17, 16–17 May 2020, Chennai, India (2020). https://doi.org/10.4108/eai.16-5-2020.2304170

4. Edquist, H., Goodridge, P., Haskel, J.: The Internet of Things and economic growth in a panel of countries. Econ. Innov. New Technol. **30**(3), 262–283 (2021). https://doi.org/10.1080/10438599.2019.1695941
5. Espinoza, H., Kling, G., McGroarty, F., O'Mahony, M., Ziouvelou, X.: Estimating the impact of the internet of things on productivity in Europe. Heliyon **6**(5)(e03935) (2020). https://doi.org/10.1016/j.heliyon.2020.e03935
6. Schiller, E., et al.: Landscape of IoT security. Comput. Sci. Rev. **44**(100467) (2022). https://doi.org/10.1016/j.cosrev.2022.100467
7. Nistor, A., Zadobrischi, E.: Analysis and estimation of economic influence of IoT and telecommunication in regional media based on evolution and electronic markets in Romania. Telecom **3**(1), 195–217 (2022). https://doi.org/10.3390/telecom3010013
8. Al-Shetwi, A.: Sustainable development of renewable energy integrated power sector: trends, environmental impacts, and recent challenges. Sci. Total Environ. **822**(153645) (2022). https://doi.org/10.1016/j.scitotenv.2022.153645
9. Manavalan, E., Jayakrishna, K.: A review of Internet of Things (IoT) embedded sustainable supply chain for industry 4.0 requirements. Comput. Ind. Eng. **127**(C), 925–953 (2019). https://doi.org/10.1016/j.cie.2018.11.030
10. Chau, M.Q., et al.: Prospects of application of IoT-based advanced technologies in remanufacturing process towards sustainable development and energy-efficient use. Energy Sources Part A: Recovery, Utilization Environ. Effects **2021**(1994057), 1–25 (2021). https://doi.org/10.1080/15567036.2021.1994057
11. Memić, B., Hasković-D||zubur, A., Avdagić-Golub, E.: Green IoT: sustainability environment and technologies. Sci. Eng. Technol. **2**(1), 24–29 (2022). https://doi.org/10.54327/set2022/v2.i1.25
12. Rosca, M.I., Nicolae, C., Sanda, E., Madan, A.: Internet of Things (IoT) and Sustainability. In: Proceedings of the 7th BASIQ International Conference on New Trends in Sustainable Business and Consumption (BASIQ 2021), pp. 346–352, 3–5 June 2021, Foggia, Italy (2021). https://doi.org/10.24818/BASIQ/2021/07/044
13. Boto-Álvarez, A., García-Fernández, R.: Implementation of the 2030 agenda sustainable development goals in Spain. Sustainability **12**(6)(2546) (2020). https://doi.org/10.3390/su12062546
14. Norouzi, N.: Mathematics of the circular economics: a case study for the MENA region. In: Handbook of Research on Building Inclusive Global Knowledge Societies for Sustainable Development, 1st edition, 143–165, IGI Global, Pensivania, United States of America (2022). https://doi.org/10.4018/978-1-6684-5109-0.ch007
15. Zara, C., Ramkumar, S.: Circular economy and default risk. J. Financ. Manage. Markets Institutions **2022**(2250001), 1–24 (2022). https://doi.org/10.1142/S2282717X22500013
16. Martínez, I., et al.: Internet of Things (IoT) as sustainable development goals (SDG) enabling technology towards smart readiness indicators (SRI) for university buildings. Sustainability **13**(14)(7647) (2021). https://doi.org/10.3390/su13147647
17. Rejeb, A., Suhaiza, Z., Rejeb, K., Seuring, S., Treiblmaier, H.: The Internet of Things and the circular economy: a systematic literature review and research agenda. J. Cleaner Prod. **350**(131439) (2022). https://doi.org/10.1016/j.jclepro.2022.131439
18. Chu, X., Nazir, S., Wang, K., Leng, Z., Khalil, W.: Big data and its V's with IoT to develop sustainability. Sci. Programm. **2021**(3780594) (2021). https://doi.org/10.1155/2021/3780594

19. Sun, X., Wang, X.: Modeling and analyzing the impact of the internet of things-based industry 4.0 on circular economy practices for sustainable development: evidence from the food processing industry of China. Front. Psychol. **13**(866361) (2022). https://doi.org/10.3389/fpsyg.2022.866361
20. Suresh, T., Murugan, A.,: Strategy for data center optimization: improve data center capability to meet business opportunities. In: 2nd International Conference on IoT in Social, Mobile, Analytics and Cloud (I-SMAC 2018), 30–31 August 2018, Palladam, India (2018). https://doi.org/10.1109/I-SMAC.2018.8653702
21. Sadrosadati, M., Mirhosseini, A., Akbarzadeh, N., Aghilinasab, H., Sarbazi-Azad, H.: Chapter Two - an efficient DVS scheme for on-chip networks. Adv. Comput. **124**, 21–43 (2022). https://doi.org/10.1016/bs.adcom.2021.09.002
22. Costa, S., Dalai, M., Pasotti, A.: A tour problem on a toroidal board. Australas. J. Comb. **76**(1), 183–207 (2020). https://ajc.maths.uq.edu.au/pdf/76/ajc_v76_p183.pdf
23. Miao, L., Zhang, S., Li, R.H., Yang, W.: Structure fault tolerance of k-ary n-cube networks. Theor. Comput. Sci. **795**(C), 213–218 (2019). https://doi.org/10.1016/j.tcs.2019.06.013
24. Xie, Y., Liang, J., Yin, W., Li, C.: The properties and t/s-diagnosability of k-ary n-cube networks. J. Supercomput. **78**(5), 7038–7057 (2022). https://doi.org/10.1007/s11227-021-04155-y
25. Yan, H.: News and public opinion multioutput IoT intelligent modeling and popularity big data analysis and prediction. Comput. Intell. Neurosci. **2022**(3567697) (2022). https://doi.org/10.1155/2022/3567697
26. Lv, Z., Qiao, L., Verma, S., Kavita, A.: AI-enabled IoT-edge data analytics for connected living. ACM Trans. Internet Technol. **21**(4)(104) (2021). https://doi.org/10.1145/3421510
27. Bergstra, J.A., Middleburg, C.A.: Using Hoare logic in a process algebra setting. Fund. Inform. **179**(4), 321–344 (2021). https://doi.org/10.3233/FI-2021-2026
28. Fokkink, W.: Modelling Distributed Systems. Springer-Verlag, Berlin, 2nd ed., 1–174, ISBN: 978-3-540-73938-8, Heidelberg, Germany (2017). https://doi.org/10.1007/978-3-540-73938-8
29. Roig, P.J., Alcaraz, S., Gilly, K., Bernad, C., Filiposka, S.: De Bruijn-based and k-ary n-cube-based algebraic models in fog environments. Commun. Comput. Inf. Sci. (CCIS) **1521**, 126–141 (2022). https://doi.org/10.1007/978-3-031-04206-5_10
30. Groote, J.F., Mousavi, M.R.: Modeling and Analysis of Communicating Systems. MIT Press, 1st ed., 1–392, ISBN: 978-0-262-02771-7, Cambridge, Massachusetts, USA (2014). https://mitpress.mit.edu/books/modeling-and-analysis-communicating-systems
31. Fokkink, W.: Introduction to Process Algebra. Springer-Verlag, Berlin, 2nd ed., 1–151, ISBN: 978-3-662-04293-9, Heidelberg, Germany (2007). https://doi.org/10.1007/978-3-662-04293-9
32. Erra, R.: Implementation of a Hardware Algorithm for Integer Division. M.S. thesis, Elect. Eng., Fac. Graduate College Oklahoma State Univ., USA (2019). https://hdl.handle.net/11244/323391
33. Ahn, Y., Lee, Y., Lee, G.: Power and time efficient IP lookup table design using partitioned TCAMs. Circ. Syst. **2013**(4), 299–303 (2013). https://doi.org/10.4236/cs.2013.43041

Towards Energy-Aware Management of Shared Printers

Antreas Kasiotis, Chinomnso Ekwedike, and Matthew Forshaw(✉)

School of Computing, Newcastle University, Newcastle upon Tyne NE1 7RU, UK
{chinomnso.ekwedike,matthew.forshaw}@newcastle.ac.uk

Abstract. The energy efficiency of IT estates face increasing scrutiny in terms of consumption and CO_2 emissions. Existing research efforts predominantly consider energy efficiency at the server and datacentre level. Meanwhile, the energy aware operation of multi-function printer devices has received relatively little attention. Printing incurs significant energy consumption and cost for organisations, and is estimated to be responsible for 10–16% of ICT-related energy consumption within higher education. We present early efforts to develop energy- and occupany-aware management policies for printers. We evaluate our work against an institutional trace of print jobs and building occupancy. We demonstrate the potential to optimise operation by minimising operating time by 4.8%, reducing the number of on-off state transitions by 35%.

Keywords: Energy Efficiency · Printing · Simulation

1 Introduction

Printing is estimated to account for 10–16% of ICT related energy electricity consumption within higher education [27]. Meanwhile, the parameterisation of existing energy saving mechanisms is shown to be problematic [20,29], limiting their efficacy. IT estate deployments have reported that printers may be powered down for only 15–30% of the time [26]. Hence, there is clearly demand for improved, data-driven approaches to handling the energy consumption of printers.

Prior efforts to promote sustainability in printing have primarily focused on informing and altering user behaviour to reduce usage of consumables. Despite clear opportunities for significant reductions in cost and environmental impact, few have considered energy-efficient management policies for shared printing.

This paper aims to demonstrate the feasibility of energy- and occupancy-aware management strategies for printers in shared environments. We seek to develop approaches to predict future demand and optimise their operation and energy consumption by transitioning between low-power and active states.

The remainder of this paper is organised as follows. Section 2 introduces background and related work. Section 3 introduces our system model and experimental design. We present our results and findings in Sect. 4. In Sect. 5 we conclude and offer suggestions for related areas of future research.

M. Forshaw et al. (Eds.): PASM 2022, CCIS 1786, pp. 93–104, 2023.
https://doi.org/10.1007/978-3-031-44053-3_6

2 Background and Related Work

Occupancy-based energy consumption forecasting has long been an extensively researched topic in the field of energy management [40]. For example, occupancy data has been used in energy saving policies for multi-use clusters [7,18,33,35–37] serving students and volunteer computing systems [19], workflow scheduling [34] and checkpointing strategies [17] and VM resource management [1–3].

Energy management policies involve either predictive or stochastic approaches [31]. In case of printers, dynamic timeout policies are typically proposed which use the distribution of historical usage data [45]. However, few leverage occupancy data [40], despite its prior use in cluster management [16,37]. Stochastic techniques commonly model the time between user requests as Markov decision processes to find the optimal waiting time before a device enters sleep mode [5,15]. Several studies utilised machine learning methods for devising dynamic policies such as classification, regression, and reinforcement learning [11,41].

Ciriza *et al.* [12] develop a statistical model for the optimisation of power consumption of printers by determining the optimal printer timeout period. Applied to a data set comprising 2,320 jobs collected over seven months, the developed approaches achieving 17.8–19.2% reduction in energy consumption compared to a baseline timeout period of 30 min. However, these savings come at the cost of a 296–326% increase in the number of shutdown and wake-up transitions, posing significant implications on printer reliability.

Andreoli *et al.* [4] employ a clustering approach to analyse usage data from an infrastructure of shared printer devices, aiming to discover communities of device usage based on physical or virtual location.

Stefanek *et al.* [42] present energy consumption data of a number of office devices collected over thirty days, including a single centrally-managed shared printer within a university. They observe this printer consumed 108.4 kWh in March 2012, spending a large proportion of its time in an idle state, transitioning into a very high power mode to serve print jobs.

Other works have focused on user behaviour, Grasso *et al.* [21] present the Personal Assessment Tool (PAT), a tool designed to assist users optimise their use of printing resources, and provide visibility into the user's print behaviour compared to that of their colleagues, and relative to organisational goals. Willamowski *et al.* [43] present the qualitative results of a user study of the PAT tool.

Finally a number of commercial print management solutions exist to reduce the impact of printing by provide accountability for print resources, and identify opportunities for cost and energy savings [22] or optimising documents to lower toner consumption without impacting on print quality [39]. Manufacturers commonly leverage sustainability claims in marketing materials, a number of whom offer tools to measure the environmental impact of their products [25,44].

We focus primarily on the operational environmental impact of printers. Considerable additional environmental impact results from procurement and end-of-life disposal of devices, as well as the use of consumables including paper and toner/ink cartridges. A survey and critical evaluation of life-cycle analysis of the environmental impact of print resources is presented in [6].

3 System Model

In this section we outline our system model for shared print infrastructures, and approach to power consumption modelling of print resources.

Jobs: We consider two classes of job; namely *colour* and *black and white*. We also consider properties of the document e.g. duplex or simplex printing.

Printer: We consider *colour* or *black and white only* laser printers, with all colour printers assumed to be able to also process black and white jobs. While devices of the same manufacturer and model may exhibit differences due to manufacturing variation, device age and usage profile, and the impact of replacement parts and servicing, we assume heterogeneous performance and power profiles.

We consider laser and multifunction (MFD) printers with users specifying a destination at the time of printing (*"push printing"*). Conversely, some organisations submit jobs to a print server, and are processed only when the user arrives and authenticates at a print station (*"pull printing"*). This offers significant advantages in flexibility to users and mitigates waste due to uncollected printouts, but may impact users negatively for long-running jobs where users will have to endure a delay while the job prints.

We model the time taken to print a job j on printer p (in seconds) as follows;

$$D_{j,p} = \begin{cases} s_j d_j \times \frac{60}{ppm_{p,c_j,f_j}} + \frac{60}{ppm_{p,0,0}} & \text{if } hs_p \\ s_j d_j \times \frac{60}{ppm_{p,c_j,f_j}} & \text{otherwise} \end{cases} \tag{1}$$

where s_j is the number of sides in the document printed as part of j, d_j is the number of copies of the document, $c_j \in [0,1]$ denotes whether the print job is black and white ($c_j = 0$) or colour ($c_j = 1$), $f_j \in [0,1]$ represents the format of print job j - whether simplex ($f_j = 0$) or duplex ($f_j = 1$), $ppm_{p,m}$ is the print speed (measured in pages per minute) for printer p in mode m (either black and white or colour) in format f (either simplex or duplex), and $hs_p \in [0,1]$ states whether the job prints with a header sheet ($hs_p = 1$) or not ($hs_p = 0$).

Group: We consider logical printer *groups*, comprising one or more print device serving a particular physical location and/or subset of users. We assume groups of printers to be homogeneous in their capabilities, but power and performance may vary within each group. One or more groups of printers may occupy a single physical location; for example a student cluster with both black and white and colour printer devices will be modelled as two distinct groups.

Users: In traditional office environments users generally have dedicated machines and work locations so their preference in printer is predictable. We attribute this behaviour to 'staff' users, where the number of users per printer is also relatively small to minimise the disruption of moving from desktop to printer. Meanwhile, our scenario of an academic institution more closely represents that of a *'hot desking'* environment for student users, with printing resources shared between a much larger number of users. While student users may exhibit preferential decisions on where they work (e.g. a particular cluster in the Students' Union building or Library, or a particular departmental cluster), we observe users submitting print jobs from multiple locations across campus.

3.1 Printer Model

We model a printer as belonging in one of three states; *Sleep*, *Idle* or *Active/ Print*. Transitions between these states are shown in Fig. 1. Transitions between Sleep and Idle states may take place according to a power managemment policy, such as EnergyStar, or through manual intervention.

Sleep The device is unavailable to service incoming requests immediately. A warm-up period of up to several minutes is required to tranistion to Idle or Active/Print states, during which high power draw is observed while components are pre-heated. Policies should minimise the number of transitions between Sleep and Idle states to minimise mechanical wear-and-tear.

Idle The device is ready to service incoming jobs immediately. Commonly print devices may experience a relatively high 'base' energy consumption, relative to active operation. Therefore, minimising the amount of time a printer is in Idle mode is advantageous in reducing energy consumption.

Active/Print The device is actively serving a print job, the state in which energy consumption is greatest.

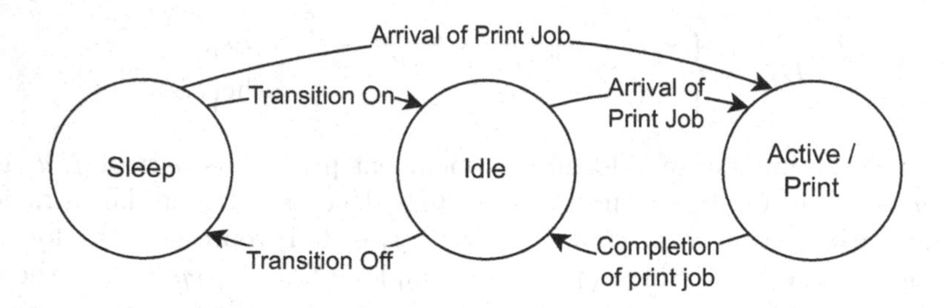

Fig. 1. State transition model for printer device.

3.2 The Data

Our study uses five years of data concerning printer usage and building occupancy. Printers serve computer clusters, and as our proxy for occupancy we use logs of user log-ins and log-outs within the computer clusters the printers serve.

Our *jobs* dataset holds details of the print requests received, including a timestamp for their arrival, the user, the print location chosen, and metadata concerning the job itself (including number of pages, colour/black and white, page count, number of copies, duplex/simplex).

Our *user* dataset includes tuples containing login and logout timestamps for individual users, a unique ID for the user and the computer they logged onto.

Figure 1A shows the number of print jobs received per day. Here we can see clear seasonality throughout the year, and within the week. Additionally, we observe clear differences between term-time and holiday periods.

Imbalance in Printer Utilisation When we look at Fig. 1C, we can see an imbalance in the number of print jobs arriving at individual printers. Printer 96 experiences $\sim 10\times$ the number of requests as the second busiest printer. This may be due to several factors:

Location: individual locations may receive more or less footfall than others. For example, we expect printers serving ground-floor computer clusters to receive more traffic than basement-level computer clusters which are governed by an additional layer of swipe card access.

Job distribution strategy: Three printer selection strategies are commonly used in practice, determining which of a group of printers is selected to service an arriving job. These are *random selection*, *round-robin*, and *primary-secondary*. The random selection and round-robin policies distribute load roughly evenly across available resources, while configuring printers in a primary-secondary configuration aims to load one printer more heavily, using others only to service high-levels of demand. The benefit of this approach is to reduce the likelihood of multiple printers in a group failing or running low on consumables at the same time. Furthermore, primary-secondary is beneficial in terms of energy consumption, while other policies incur much greater overheads.

Correlation Between Print Usage and Occupancy Fig. 2B demonstrates the link between occupancy and the number of print jobs across 2019, encouraging us that occupancy-aware print strategies may provide improved performance.

We conduct Spearman correlation tests on the print job demand and occupancy, at an hourly level, for each printer and individual computer cluster. The findings are present in Table 1. We see strongest correlation at the Whole Building level, and for clusters located in prominent thoroughfares.

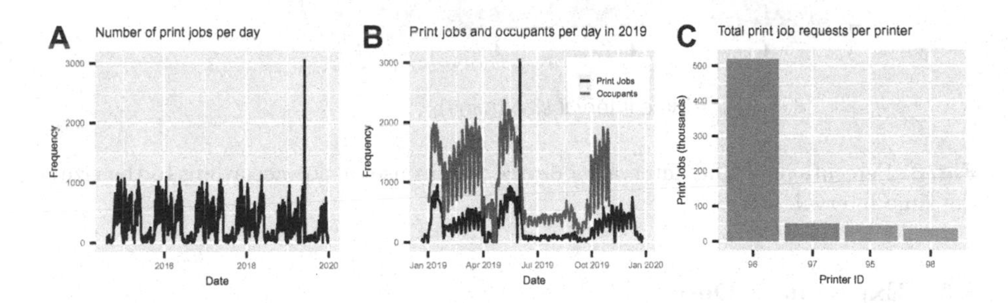

Fig. 2. In Plot A are the print job requests per day, in Plot B are both the print job and the cluster occupancy per day for 2019 and in Plot C are the total print job requests of each printer.

Table 1. Spearman correlation of Print job loads and Occupancy per hour, for every printer and cluster in the Library.

	All Print Jobs	Printer			
		95	96	97	98
EasyAccess	0.5771582	0.4446143	0.5688153	0.3828360	0.3337781
Building - Assorted	0.6493562	0.4996929	0.6426945	0.4264972	0.3680380
1st Floor - Cluster A	0.6912396	0.5376874	0.6767327	0.5001407	0.5327141
1st Floor - Cluster B	0.2415291	0.1922950	0.2389392	0.1506368	0.1294193
1st Floor - Cluster C	0.7291480	0.5540212	0.7131130	0.5696420	0.4520412
2nd Floor	0.8412543	0.6232093	0.8264987	0.5169003	0.4444037
2nd Floor - Cluster A	0.6293317	0.5104989	0.6258062	0.4576644	0.4346367
3rd Floor	0.8356925	0.6129554	0.8239115	0.4997804	0.4346735
4th Floor	0.8471182	0.6189153	0.8367839	0.5007733	0.4432568
Cluster E	0.0244190	0.0461315	0.0159339	0.0248476	0.0704942
PopUp	0.8093512	0.6027981	0.7960379	0.5155563	0.4593505
Whole Building	0.8509266	0.6258818	0.8377877	0.5163469	0.4497196

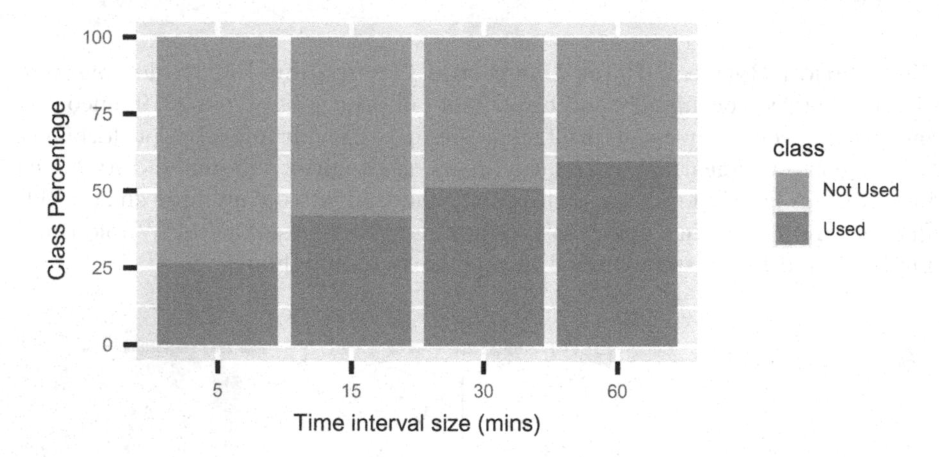

Fig. 3. Percentage of time intervals a device was in use or not according to the size of the time interval.

3.3 Experiment Design

Problem Definition. We formulate this problem as a classification task, dividing our timeseries into discrete interval lengths for prediction. For each interval we predict whether the interval will have zero or non-zero print job count. Figure 3 shows, for each interval length, the proportion of zero- and non-zero job count intervals within our dataset. The impact of our choice of interval length can be found in Fig. 3. We find that smaller interval lengths lead to a finer level of granularity, a point beyond which predictive performance deteriorates. Our modelling adopts a train-test ratio of 90:10.

Printer for Evaluation. We select Printer 96 as the most representative printer for the remaining of our experimentation. We consider both occupancy-aware and temporal-only models and evaluate their relative performance.

Feature Engineering. We leverage the following temporal features using *lubridate* [23]; quarter, month, day of the week, day of the month, day of the year, weekend/working day, hour of the day, and minute of the hour. We further leverage domain-specific information such as whether there is a public holiday, whether the University is during term, exam or holiday season.

We also leverage real-time information on building occupancy (as measured by cluster room logins), and historical demand. Counts of print jobs per interval were calculated using the *Timetk* library [14], and cumulative occupancy was calculated with help of the *patientcounter* library [32].

We evaluate multiple models, to evaluate whether occupancy data allows us to improve on models which only use temporal features.

For each dataset (5, 10, 30, 60 min) we perform feature selection using *Recursive Feature Elimination (RFE)* [9] using the *caret* library [28]. The best predictor variables for each model are shown below:

- $60m.jobs \sim occupants + yday + quarter + months + day + weekdays + weekend + week + hours$
- $30m.jobs \sim occupants + yday + quarter + months + day + weekdays + weekend + week + hours + minute$
- $15m.jobs \sim occupants + yday + quarter + months + day + weekdays + weekend + week + hours + minute$
- $5m.jobs \sim occupants + yday + months + day + weekdays + weekend + week + hours$

Classifiers. We consider standard classification and regression algorithms included in the *caret, randomForest, and e1071* libraries [28,30,38]. In this paper we present our results for Support Vector Machines (SVM) [13], the Random Forest (RF) [24], the package of Generalized Linear Models (glmnet), and eXtreme Gradient Boost (xgbLinear) [10].

4 Results

We evaluate the performance of our classiers with respect to balanced accuracy, which considers both the true positive and true negative rates, mitigating bias arising due to class imbalance [8]. Figure 4 depicts the balanced test accuracy of each classifier for each of the four datasets. In the case of the 60-minute interval, the best classifier was occupancy-aware *glmnet*, with a test accuracy of 89.49%.

For the 30-minute interval, the best classifier was again *glmnet*, trained with only temporal features, with a test accuracy of 87.82%. The set with a 15-minute interval, achieved a test accuracy of 83.7% with an occupancy-aware *SVM* classifier. Lastly, for the 5-minute interval, the best classifier was occupancy-aware *xgbLinear*, achieving a test accuracy of 62.61%. Our best performing classifier was glmnet trained on the 60-minute dataset with temporal and occupancy features, achieving almost a 90% balanced accuracy.

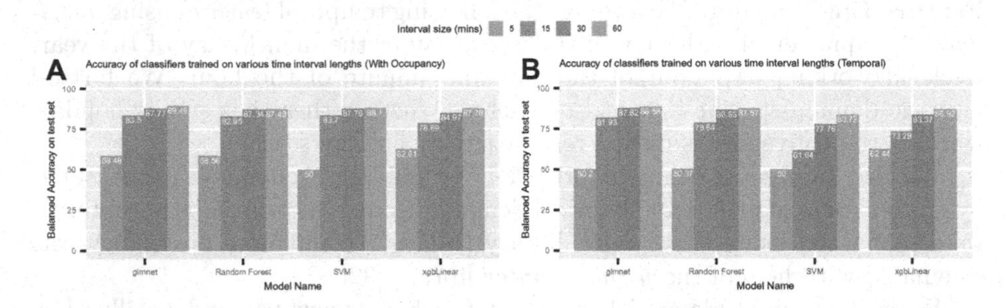

Fig. 4. Balanced accuracy of classifiers trained with occupancy (Plot A) and only temporal (Plot B) features.

The average balanced test accuracy of models trained with both temporal and environmental attributes on all four datasets came out to 77.39%, 79.08%, 79.81%, and 78.51% for the SVM, Random Forest, glmnet, and xgbLinear respectively. The same models that were trained with only temporal variables produced average balanced accuracies of 68.28%, 75.80%, 77.13%, and 76.50%. With the exception of the *glmnet* model for the 30-minute intervals, we see that occupancy-aware models perform best. Our classifiers perform best at courser granularity. Performance degrades significantly from the 15-min to 5-min case.

Table 2. Performance evaluation of Temporal and Occupancy-aware models, with respect to balanced training and test accuracy, and training time.

Classifiers	Bal. Train Acc. (%)				Bal. Test Acc. (%)				Training Time (Secs)			
	60	30	15	5	60	30	15	5	60	30	15	5
Temporal												
SVM	81.74	79.69	76.07	73.92	83.72	77.76	12.34	50	0.9	4.4	20.3	102.9
RF	93.73	96.51	94.20	78.17	87.57	86.63	0.45	50.37	1.0	2.6	5.7	11.1
glmnet	83.13	80.85	77.82	73.41	88.58	87.82	30.55	50.2	21.8	41.9	82.6	292.5
xgbLinear	92.25	89.34	85.33	77.44	86.92	87.34	30.55	62.44	223.7	338.3	617.1	1435.3
Temporal and Occupancy												
SVM	83.91	81.33	79.24	73.92	88.1	87.76	12.34	50	1.0	4.5	21.9	135.3
RF	99.9	99.38	98.08	80.47	87.48	87.34	0.45	58.56	4.30	3.1	6.7	14.5
glmnet	84.6	82.18	79.25	73.97	89.49	87.77	30.55	58.48	25.4	48.0	94.6	311.7
xgbLinear	93.44	89.66	86	77.79	87.78	84.97	30.55	62.61	227.0	343.1	623.7	1440.0

Table 2 shows the training time varies across classifiers, with the slowest being the xgbLinear that was trained on the 5-minute dataset and the fastest being the SVM that was trained on the 60-minute dataset. It is also apparent that the smaller the interval size of a dataset the longer it takes to train a model. When looking at the training accuracies of the models it seems that some models tend to overfit. It is clear from the tables above that *Random Forest* is producing a much better training accuracy when compared to its test accuracy. This is the case for all classifiers that were trained with a smaller time interval. *Random Forest* is the quickest and the *xgbLinear* is the slowest on average. A great balance is found in the *glmnet* which performed the best and was relatively quick to train.

4.1 Printer Policy Optimization

We evaluate, through trace-driven simulation, the likely impact of our best-performing classifiers. We evaluate the proportion of time printers spend awake or asleep, and the proportion of jobs arriving at a printer which is awake.

We simulate a power management policy, using the classifiers to govern decisions to transition the printers between on and off states. Our strategies operate such that printers remain in their state until the end of the interval. Table 3 shows the performance for each of our classifiers. With the exception of our five-minute policy, all classifiers reduce the number of state transitions, and a comparable proportion of print jobs arrive to an available printer.

The five minute policy is the most aggressive in its power management, with the printers only awake for 25.68% of the duration. This achieves a ~54% reduction in power consumption. However, the printers are only available to 73.2% of jobs at the point of arrival, and would impose delays to jobs arriving when the print devices are asleep. The five minute policy also incurs 4.64× the number of state transitions, increasing the likelihood of mechanical failures.

Table 3. Printer Policy simulation results.

	Interval				Existing Policy
	60	30	15	5	
State Transitions (Count)	77	130	264	1877	404.00
Jobs reaching Available printer (%)	99.17	98.35	96.34	73.20	94.10
Proportion of Time Awake (%)	63.47	56.78	50.77	25.68	55.57

Our results highlight the importance of correctly calibrating the interval. We recognise this will be dependent on the arrival process for print jobs. In our experiments, the 15-minute policy balances the trade-off between transitions and operating cost. Compared to the existing policy was found to have 35% fewer transitions, 2.24% more printer availability and 4.8% less printer operation time.

5 Conclusion and Future Work

Printing poses a significant financial and environmental impact for many organisations. In this paper we have motivated the need for energy-efficient operating policies for shared printing environments, highlighted key issues in current approaches to printer management. We have demonstrated the potential for demand prediction models using occupancy data to optimise printer operation. We have been able to reduce operating time by 4.8%, reducing the number of on-off state transitions by 35%, while maintaining quality of service.

Future work could consider the potential of hierarchical models as a principled approach to considering printer- and building-level effects. Similarly, time-to-event models offer an alternative approach to model uncertainty of estimates. User-level models of print propensity were outside the scope of this work, but have the potential to offer increased predictive performance.

References

1. Alrajeh, O., Forshaw, M., Thomas, N.: Virtual machine live migration in trace-driven energy-aware simulation of high-throughput computing systems. Sustain. Comput.: Inform. Syst. (2020)
2. Alrajeh, O., Forshaw, M., McGough, A.S., Thomas, N.: Simulation of virtual machine live migration in high throughput computing environments. In: 2018 IEEE/ACM 22nd International Symposium on Distributed Simulation and Real Time Applications (DS-RT), pp. 1–8. IEEE (2018)
3. Alrajeh, O., Forshaw, M., Thomas, N.: Using virtual machine live migration in trace-driven energy-aware simulation of high-throughput computing systems. Sustain. Comput.: Inform. Syst. **29**, 100468 (2021)
4. Andreoli, J.-M., Bouchard, G.: Probabilistic latent clustering of device usage. In: Famili, A.F., Kok, J.N., Peña, J.M., Siebes, A., Feelders, A. (eds.) IDA 2005. LNCS, vol. 3646, pp. 1–11. Springer, Heidelberg (2005). https://doi.org/10.1007/11552253_1
5. Benini, L., Bogliolo, A., De Micheli, G.: A survey of design techniques for system-level dynamic power management. IEEE Trans. Very Large Scale Integration (VLSI) Syst. **8**(3), 299–316 (2000). https://doi.org/10.1109/92.845896
6. Bousquin, J., Esterman, M., Rothenberg, S.: Life cycle analysis in the printing industry: a review. In: NIP, pp. 709–715 (2011)
7. Bradley, J.T., Forshaw, M., Stefanek, A., Thomas, N.: Time-inhomogeneous population models of a cycle-stealing distributed system. UKPEW **2013**, 8 (2013)
8. Brodersen, K.H., Ong, C.S., Stephan, K.E., Buhmann, J.M.: The balanced accuracy and its posterior distribution. In: 2010 20th International Conference on Pattern Recognition, pp. 3121–3124 (2010). https://doi.org/10.1109/ICPR.2010.764
9. Bulut, O.: Effective Feature Selection: recursive feature elimination using R (2021). https://towardsdatascience.com/effective-feature-selection-recursive-feature-elimination-using-r-148ff998e4f7
10. Chen, T., Guestrin, C.: XGBoost: a scalable tree boosting system. In: Proceedings of the 22nd ACM SIGKDD International Conference on Knowledge Discovery and Data Mining, pp. 785–794. KDD '16, ACM (2016). https://doi.org/10.1145/2939672.2939785

11. Chiming Chang, Paul Armand Verhaegen, J.R.D.: A comparison of classifiers for intelligent machine usage prediction. In: 2014 International Conference on Intelligent Environments, pp. 198–201 (2014). https://doi.org/10.1109/IE.2014.36
12. Ciriza, V., Donini, L., Durand, J.B., Girard, S.: A statistical model for optimizing power consumption of printers. In: Presentation during a joint meeting of the Statistical Society of Canada & the Société Française de Statistique (2008)
13. Cortes, C., Vapnik, V.: Support-vector networks. Mach. Learn. **20**(3), 273–297 (1995)
14. Dancho, M., Vaughan, D.: timetk: a tool kit for working with time series in r (2022). https://CRAN.R-project.org/package=timetk
15. Durand, J.B., Girard, S., Ciriza, V., Donini, L.: Optimization of power consumption and device availability based on point process modelling of the request sequence. J. Royal Stat. Soc. Ser. C (Appl. Stat.) **62**(2), 151–165 (2013). http://www.jstor.org/stable/24771799
16. Forshaw, M., McGough, A.S.: Flipping the priority: effects of prioritising HTC jobs on energy consumption in a multi-use cluster. Social-Informatics and Telecommunications., Institute for Computer Sciences (2015)
17. Forshaw, M., McGough, A.S., Thomas, N.: On energy-efficient checkpointing in high-throughput cycle-stealing distributed systems. In: 3rd International Conference on Smart Grids and Green IT Systems (SMARTGREENS) 2014 (2014)
18. Forshaw, M., McGough, A.S., Thomas, N.: Energy-efficient checkpointing in high-throughput cycle-stealing distributed systems. Electron. Notes Theor. Comput. Sci. **310**, 65–90 (2015)
19. Forshaw, M., McGough, A.S., Thomas, N.: HTC-Sim: a trace-driven simulation framework for energy consumption in high-throughput computing systems. Concurrency Comput.: Pract. Exper. **28**(12), 3260–3290 (2016)
20. Gingade, G., Chen, W., Lu, Y.H., Allebach, J., Gutierrez-Vazquez, H.I.: Hybrid power management for office equipment, **22**(1) (2016).https://doi.org/10.1145/2910582
21. Grasso, A., Willamowski, J., Ciriza, V., Hoppenot, Y.: The personal assessment tool: a system providing environmental feedback to users of shared printers for providing environmental feedback. In: ICMLA, pp. 704–709. IEEE (2010)
22. GreenPrint: GreenPrint (Homepage). http://www.printgreener.com/
23. Grolemund, G., Wickham, H.: Dates and times made easy with {lubridate} 40 (2011). https://www.jstatsoft.org/v40/i03/
24. Ho, T.K.: Random decision forests. In: Proceedings of 3rd International Conference on Document Analysis and Recognition, vol. 1, pp. 278–282. IEEE (1995)
25. HP: Carbon Footprint Calculator for Printing. http://www.hp.com/large/ipg/ecological-printing-solutions/
26. James, P., Hopkinson, L.: Energy efficient printing and imaging in further and higher education. A Best Practice Review prepared for the Joint Information Services Committee (JISC) (2008)
27. James, P., Hopkinson, L.: Results of the 2008 susteit survey. Environ. Manage. **50**(27), 25 (2008)
28. Kuhn, M.: caret: Classification and regression training (2022). https://CRAN.R-project.org/package=caret
29. KYOCERA Document solutions Inc.: Waking an electronic device, such as a printer, from sleep mode based on a user policy and proximity (2022)
30. Liaw, A., Wiener, M.: Classification and regression by randomforest **2**, 18–22 (2002). https://CRAN.R-project.org/doc/Rnews/

31. Lu, Y.H., Chung, E.Y., Šimunić, T., Benini, L., De Micheli, G.: Quantitative comparison of power management algorithms. In: Proceedings of the Conference on Design, Automation and Test in Europe, pp. 20–26. DATE '00 (2000). https://doi.org/10.1145/343647.343688
32. MacKintosh, J.: patientcounter: Count hospital patients quickly (2022). https://github.com/johnmackintosh/patientcounter
33. McGough, A.S., Forshaw, M.: Reduction of wasted energy in a volunteer computing system through reinforcement learning. Sustain. Comput.: Inform. Syst. **4**(4), 262–275 (2014)
34. McGough, A.S., Forshaw, M.: Energy-aware simulation of workflow execution in high throughput computing systems (2016)
35. McGough, A.S., Forshaw, M.: Evaluation of energy consumption of replicated tasks in a volunteer computing environment. In: Companion of the 2018 ACM/SPEC International Conference on Performance Engineering, pp. 85–90. ACM (2018)
36. McGough, A.S., Forshaw, M., Gerrard, C., Wheater, S., Allen, B., Robinson, P.: Comparison of a cost-effective virtual cloud cluster with an existing campus cluster. Future Gener. Comput. Syst. **41**, 65–78 (2014)
37. McGough, A.S., Forshaw, M., Gerrard, C., Robinson, P., Wheater, S.: Analysis of power-saving techniques over a large multi-use cluster with variable workload. Concurrency Comput.: Pract. Exper. **25**(18), 2501–2522 (2013)
38. Meyer, D., Dimitriadou, E., Hornik, K., Weingessel, A., Leisch, F.: e1071: Misc functions of the department of statistics, probability theory group (formerly: E1071), tu wien (2022). https://CRAN.R-project.org/package=e1071
39. Preton Ltd: Preton PretonSaver. http://www.preton.com/pretonsaver.asp
40. Soetedjo, Aryuanto, S.: Modeling of occupancy-based energy consumption in a campus building using embedded devices and IoT technology. Electronics **10**(18) (2021). https://doi.org/10.3390/electronics10182307
41. Spiliotis, E., Makridakis, S., Semenoglou, A.-A., Assimakopoulos, V.: Comparison of statistical and machine learning methods for daily SKU demand forecasting. Oper. Res. **22**(3), 3037–3061 (2020). https://doi.org/10.1007/s12351-020-00605-2
42. Stefanek, A., Harder, U., Bradley, J.T.: Energy consumption in the office. In: Tribastone, M., Gilmore, S. (eds.) EPEW 2012. LNCS, vol. 7587, pp. 224–236. Springer, Heidelberg (2013). https://doi.org/10.1007/978-3-642-36781-6_16
43. Willamowski, J.K., Hoppenot, Y., Grasso, A.: Promoting sustainable print behavior. In: CHI'13, pp. 1437–1442. ACM (2013)
44. Xerox Corporation: Xerox Sustainability Calculator. http://www.consulting.xerox.com/flash/thoughtleaders/suscalc/xeroxCalc.html
45. Xerox Corporation: Patent US20140181552A1: multi-mode device power-saving optimization. https://patents.google.com/patent/US20140181552 (2009)

Modelling Performance and Fairness of Frame Bursting in IEEE 802.11n Using PEPA

Choman Abdullah[1] and Nigel Thomas[2(✉)]

[1] College of Education, University of Sulaimani, Sulaymaniyah, Iraq
choman.abdullah@univsul.edu.iq
[2] School of Computing, Newcastle University, Newcastle upon Tyne, UK
nigel.thomas@newcastle.ac.uk

Abstract. IEEE 802.11n is a relatively inexpensive high throughput WLAN standard, which has been used in many wireless devices. The achievable capacity increases with 802.11n MAC and PHY layer enhancements. This paper presents novel performance models of frame bursting in 802.11n, specified and analysed using Markovian process algebra PEPA. We investigate performance fairness through channel utilisation and throughput, considering the frame bursting length in data transmission, by studying different communicating pairs of nodes within two scenarios in a restricted network topology. We explore the potential to reduce unfairness by allowing affected nodes to have longer burst lengths.

1 Introduction

IEEE 802.11 is a dominant standard of WLAN due to its low cost, high speed and easy development [1]. WLAN access adopts widely as a network medium of choice via its scalability, mobility and flexibility for roaming users. The entire urban areas can be covered by WLAN that increases pervasively to obtain network services to users without being tethered to any cable. WLAN has affected the mobility of computer networks [2], and the evolution of WLAN users has became more mobile [3].

WLAN performance is essential in making appropriate choices for the provision of infrastructure and services. Network throughput and latency are important to know whether the network can support a given level of service. Fairness is concerned with the forced variability of throughput and latency at different nodes leading to different parts of the network attaining different levels of performance. Users, devices or applications must receive a fair or reasonable share of system resources. Fairness can determine in WLAN when all nodes fairly and equally access the medium. If one or more nodes have less opportunity to access the bandwidth, it creates an unfair scenario. The system should be fair while there is no competing traffic, but if two or more nodes are competing each other to access the media, then any resolution of the contention should not be biased in favour of either party.

M. Forshaw et al. (Eds.): PASM 2022, CCIS 1786, pp. 105–131, 2023.
https://doi.org/10.1007/978-3-031-44053-3_7

In [4], fairness is affected by transmission rate and frame length. The models in [4] show short frames sent faster promoted a greater opportunity of sharing access, even under a pathologically unfair network topology. But, in practice it is not possible to simply set an arbitrarily short frame length and fast transmission as these factors also dictate the transmission range; in CSMA/CA neighbouring nodes need to be able to 'sense' a transmission in order to minimise and detect interference. Hence, wireless protocols provide only a small set of possible transmission rates with fixed, or at least minimum, frame lengths, allowing the network provider to choose an option which best fits its operating environment.

The achievable capacity has increased significantly in 802.11n, with the data rate reaching up to 600 Mbps [5]. 802.11n has five main technical improvements; MIMO, Spatial Multiplexing, Channel Bonding, Short Guard Interval and MAC layer enhancements. The MIMO "Multiple Input and Multiple Output" technology used in 802.11n to increase speed and data rates by using multiple transmitters and receivers at the same time by both sender and receiver. Spatial Multiplexing transmission technique in MIMO technology, has been employed the use of multiple antennas to transmit and receive independent and multiple data streams simultaneously, which support users to obtain maximum use of available bandwidth. Channel bonding is another technical improvement that increases throughput in 802.11n. It is a combination of two or more communication links or adjacent channels in which to increase the amount of data that can be transmitted. Guard Interval is used in communications to avoid interference among symbol transmissions from multipath effect. Most IEEE 802.11 protocols use 800 *ns* as a guard interval. However, the interval time in 802.11n becomes 400 *ns* as a Short Guard Interval; this shorter period of time for symbol transmission can be used to improve the throughput.

This research presents a novel model and performance analysis of frame bursting provision, by focusing on 802.11n MAC layer attributes of frame aggregation and block acknowledgement using PEPA. Several models are presented in a literature of restricted network topologies and study their performance and fairness, in this case varying the burst length to determine its effect. To the best of our knowledge, we are the first to study the analytical model 802.11n for channel throughput and utilisation.

The remainder of this paper is organised as follows. Basic access mechanism and 802.11n frame bursting are introduced in Sect. 2. PEPA is shown in Sect. 3. A literature review is introduced in Sect. 4. Section 5 discusses our scenarios with PEPA model. The parameters used in our models are described in Sect. 6. The results and discussions are discussed in Sect. 7. Finally, Sect. 8 concludes this study and discusses some future work.

2 Basic Access Mechanism and IEEE 802.11n Frame Bursting

Basic access mechanism in 802.11 cooperates by using one of two different modes; Point Coordination Function (*PCF*, which needs a central control object) and

Distributed Coordination Function (*DCF*, based on CSMA/CA). DCF in 802.11 is a common technique of basic access mechanism, which is used up to 802.11g [6]. The DCF specifies this mechanism and two-way handshake as two techniques for data transmission, (see, [7]). The basic access mechanism in newer protocols is similar to legacy family standards. More recent published protocols, such as 802.11n is introduced with the main enhancements in PHY and MAC layer based on the foundation of 802.11a/b/g/e protocols. These improvements provide a higher performance in 802.11n by raising the data rates in PHY layer and hence increasing MAC layer efficiency, [8,9].

MAC layer improvement and frame bursting as main achievable enhancements in 802.11n are support sender to transmit several frames simultaneously, during a limited duration called Transmission Opportunity (TXOP). Large number of frames can be sent in 802.11n via a medium by reducing the Inter-Frame Spacing time, in which the throughput performance and efficiency of 802.11n are significantly enhanced by frame bursting method. If a node attempts to use the medium for sending frames, then it sends a burst of frames rather than a single frame (multiple frames have been merged into one aggregation). Hence, the medium will be occupied for longer. An Acknowledgement (*ACK*) technique replaced by a Block Acknowledgement (*BACK* or *BA*) to acknowledge many received frames and reverse direction mechanism rather than individual acknowledge following every single frame of sequences. The *BA* technique allows transmission in both directions and acknowledge multiple frames. Throughput of 802.11n has improved by MAC layer enhancements and frame aggregation mechanism, by reducing the overheads over a larger number of frame bursting.

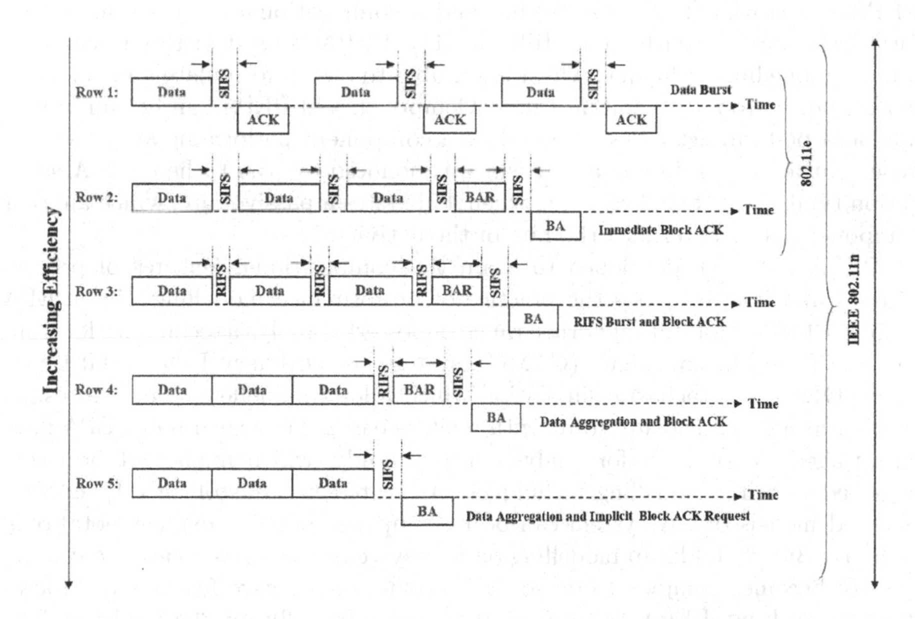

Fig. 1. IEEE 802.11n MAC layer enhancements [9,10].

802.11n frame bursting has improved step by step within five separate rows, see Fig. 1. Frame transmission in row 1 is similar to frame transmission in the old protocols, such as 802.11g. In row 1, a single frame is sent by the sender and waits to receive a single *ACK* to acknowledge each individual frames. In row 2, a Block Acknowledgement Request, *BAR*, and Block Acknowledgement, *BA*, are functioning after several frames being transmitted. This improves the protocol to reduce the number of *ACK*. Likewise, in row 3, Inter-frame Spacing time is slightly changed, and a period of time *SIFS* (Short Inter-frame Space) is replaced by *RIFS* (Reduced Inter-frame Space) between each frame. But, in row 4 the waiting time (*RIFS* and *SIFS*) are removed between frames, the main enhancement in row 4 is a concatenation of frames into frame bursting by aggregating multiple frames together; in order to achieve higher throughput. In row 4, *RIFS* is functioning after sending all frames and before *BAR*. Moreover, when the *BAR* is requested and after a period of time *SIFS*, the *BA* will be received. Finally, *BAR* removed as a last improvement in row 5, and multiple of frames transmitted as block frames. This technique will acknowledge several received frames after a period of time *SIFS*. The improvements in row 5 will lead to the higher throughput by transmitting a frame bursting with less waiting time. The basic access mechanism is considered in this paper to analyse the performance of 802.1n. We model the impact of MAC layer enhancements in different scenarios by using PEPA, as is shown in Fig. 1 (see [9,10]).

3 Performance Evaluation Process Algebra (PEPA)

PEPA is a stochastic process algebra and a compositional algebraic modelling formalism, introduced by Jane Hillston [11]. PEPA is used in this research as a main modelling technique, which is suitable to check, formulate and calculate performance properties and measures. Components in PEPA can be run in parallel and perform activities. It describes a component performing an activity of type α at activity rate r in a pair which is denoted by (α, r) when $\alpha \in \mathrm{A}$ as an action type, and rate $r \in \mathbb{R}^{+} \cup \top$, when $\top$ denotes a passive rate, which another component must determine the rate of the activity.

PEPA has been developed to exam the compositional features of process algebra and how it effects the practice of performance modelling, [11]. PEPA Eclipse Plug-in tool [12] supports a range of powerful analysis techniques for Continuous Time Markov Chain (CTMC), systems of Ordinary Differential Equations (ODEs) or stochastic simulation which allows modellers to derive results (both transient and steady state, with relative ease). The features of PEPA have encouraged us to use it for analysis in our study, which might not be easily achieved in other modelling techniques. As stated by Hillston in [11], uncomplicated models of any system can be built up without any explicit notational support. But, to build up modelling on any system, such as a computer system, quickly becomes complex to do so. Hence, the main active features of process algebras such as, PEPA, is to make tools more friendly for the modeller. The main PEPA features are shown as follows:

- **Parsimony:** PEPA is an economic tool to use with few components, accessible and flexible to modeller to utilise main features in this technique.
- **Formal definition:** This has been available by structured operational semantics provides for all expressions. The notions of equivalence can be given a formal basis for the comparison and manipulation of models and components.
- **Compositionality:** Interaction between main system and subsystems gives more ability to model the system in a proper way. Cooperation combinatory forms in PEPA is powerful as a fundamental composition. Model simplification and aggregation can be developed which are complementary to this mixture.

3.1 The Syntax and Description of PEPA

PEPA provides a useful modelling formalism to investigate properties of protocols and other systems [11]. PEPA models are specified in terms of components which interact through shared actions. Actions in PEPA have a duration, which is determined by a rate parameter of the negative exponential distribution. In shared actions a rate may be given by one or both interacting components, with the result determined by the slowest participant. In WLAN networks, components can be any node and transmission media and shared actions can be thought of as the transmission of messages from one node to another via medium. The combination of all components into a single system gives rise to labelled transition system, where the transitions between states are negative exponentially distributed actions, hence the resultant system is a CTMC. Each activity has a specific action type, and in any system a unique type can be found inside each discrete action, and countable set includes all possible types.

The structure of PEPA is not complicated, as the components and activities are the primaries in this language; small set of combinators are available which is the main combinators in PEPA: prefix, choice, co-operation and hiding. Further general information, details and structured operational semantics on PEPA can be found in [11]. The main syntax of PEPA, further details are shown the names of PEPA constructions and their intended interpretations shown as follow:

$P{::=}(\alpha,r).P \mid P \underset{L}{\bowtie} Q \mid P+Q \mid P/L \mid A$

- **Prefix,** $(\alpha, r).P$**:** This is a fundamental building block of a sequential component and basic mechanism. The process $(\alpha, r).P$ performs activity of action type α at rate r before progressing to behave as component P. A shared activity $\top$ symbol can be used as a passive participation instead of the r rate. In PEPA the actions are assumed to have a duration or delay.
- **Choice,** $P + Q$**:** Competition can be created between two or more possible processes, process $(\alpha, r).P{+}(\beta, s).Q$ any one of α will win the race (the process subsequently behaves as P) or β (the process subsequently behaves as Q).
- **Co-operation,** $P \underset{L}{\bowtie} Q$**:** The components proceed independently with any activities, and each two "co-operands" are required to operate in the co-operation to join for those activities which are specified in the co-operation set: in the process of the component of $P + Q$ represents the main system, $P \underset{L}{\bowtie} Q$ where $L = (\alpha, \beta)$ the processes P and Q must co-operate on activities

α and β but any other activities may be executed separately. The reversed compound agent theorem gives a set of sufficient conditions for a co-operation to have a product form stationary distribution. In the PEPA $P \parallel Q$ means the parallel combinator, the more concise notation $P \parallel Q$ to abbreviate for $P \underset{L}{\bowtie} Q$, where $L = \emptyset$.
- **Hiding,** P/L**:** The P/L conducts as P except that any activities of types within the set L are hidden. The process P/a hides the activity "a" from view and prevents other processes from joining with it.
- **Constant,** $A \stackrel{def}{=} P$**:** Constant A is given the behaviour of P component.

4 Literature Review

WLAN performance has been studied in a literature by examining different methods, metrics and assumptions. Many studies have been reported to reduce the different problem of 802.11 standards. As a consequence of the proliferation of protocols, performance studies considered different properties and issues, see [13,14]. Whilst the performance modelling has been employed successfully to evaluate the performance of (current and future) networking systems for many decades (see [15]). The performance characteristics of wireless networks is vital in order to obtain efficient and effective deployments. The system performance in wireless networks has been impacted by different factors, such as network topology and packet destination distribution is one of the critical impacts. These impacts discussed in [16], by studying a model to investigate the Quality of Service (QoS) performance metrics in an environment that integrates WLAN and wireless mesh. IEEE 802.16 has been investigated in [16], but we have studied how these impacts will affect the performance of 802.11 protocols.

Lee [17] used theoretical analysis to measure error-free and error-prone wireless channel with higher transmission rate in 802.11g. He studied the throughput capacity and argued that the throughput reduced if the mobility of a station increased. However, he did not consider the utilisation and did not argue for dissimilar scenario with numerous nodes. Performance degradation of 802.11g in terms of access delay studied in [18] for dissimilar nodes and throughput, by analysing collision probability, channel access delay and throughput. Ho *et al* [19] focused on WLAN throughput performance, by arguing that the best 802.11g OFDM throughput performance can be obtained when mobility is low. Also, Kanduri *et al.* [20] studied 802.11g structures in maximum data rate WLANs, as it might increase WLAN requests by users.

802.11n has replaced older protocols, and it still coexists with other protocols, such as 802.11g. The effects of coexisting of 802.11n and 802.11g is studied in [21] in wireless devices. PHY values in 802.11n increases the data rates, and MAC enhancements reduce overhead via various aspects, considered with single and multiple rates and ACK with delay ACK (see [22,23]).

802.11n provides higher speed, wide range and reliability over 802.11b/g. MIMO-based PHY and frame aggregation in MAC layer introduced in [24], and investigated the effect of MAC and PHY features in terms of fairness on adhoc networks performance. In [24] frame aggregation on the support of voice and video applications considered in wireless networks, it shows that the utilisation will be enhanced with frame aggregation of 802.11n by diminishing the protocol overheads. An analytical model for throughput, QoS provisioning in WLANs and 802.11n performance for multimedia traffics is examined in [25].

Fairness can be used to show that users, devices or applications receive a fair share of system resources. In WLAN, it can be determined when all nodes can equally access or share the medium. If one or more nodes have less opportunity to access the medium, then it creates an unfair scenario. Fairness and unfairness of 802.11 protocols have been widely studied in the literature. For example, Mohamadeou and Othman [26] have investigated the analytical evaluation of unfairness problem in WLANs. They have suggested a new mathematical model to describe the performance and impacts of WLAN dynamic behaviour.

Several studies have considered IEEE 802.11 standards in terms of the rate adaptation scheme, performance of IEEE 802.11 MAC layer, and performance metric systems. Zhai *et al* [27] attempted to "characterize the probability distribution model of the MAC layer packet service time". They have argued it has been based on deriving and creating function of the probability mass function of the inter-departure interval. Shehadeh and Chasaki have explained that to access the medium for any devices the capability and fairness are most important for reaching great effectiveness in numerous wireless devices and traffic [28].

Kloul and Valois [29] studied short and long term fairness, by establishing performance evaluation of WLAN protocol of fairness to access channel for the communicating pairs in terms of medium utilisation and throughput. Kloul and Valois have examined few scenarios to investigate the fairness of 802.11b only. Duda [30] argued the issue of unfairness, as highlighted in the above research, is a consequence of the manner in which 802.11b was implemented on early switches and that modifications made to later switches alleviate this problem. But, this does not seem to have been confirmed empirically by any researchers.

Despite the WLAN modelling progress and benefits of using PEPA to model and analyse protocols. There are very few examples in the literature where PEPA has been used to study the performance modelling of the IEEE 802.11 family, especially the latest protocol, such as 802.11n protocol. Argent-Katwala *et al.* [31] studied WLAN protocols and performance models of 802.11 in terms of its Quality of Service (QoS) based on PEPA. They argued that most of the technologies have been developed to enhance the reliability of computer networks. Moreover, they used PEPA to find properties which cannot be easy to find manually in terms of computing quantitative, passage time and increase higher probability for performance demands in wireless communication.

In [32], process algebra and PEPA are used to study DCF 802.11, by describing the handoff mechanism, quantitative analysis and channel mobility. Performance analysis of 802.11b studied in [29] by using PEPA to develop two models of network topologies, which have an effect on performance of 802.11b. They investigated an unfairness scenario in MANET, and analysed the system behaviour to measure and investigate 802.11b performance with different scenarios (three pairs scenario is one of them). In our research we have studied the three pairs scenarios, as. We present that there are uncertainty of fairness to access the channel for two pairs and unfairness for three pairs scenario in terms of medium utilisation and throughput using the same approach as [29].

5 Modelling Network and Scenarios in 802.11n with a PEPA Model

We used PEPA to investigate the highlighted different performance issues by studying the medium access in terms of channel utilisation and throughput. Two pairs and three pairs scenarios are presented to investigate the fairness in 802.11n. The "listen before talk" approach are used in each scenario, which a particular node senses the medium (attempts to transmit a frame(s) via a wireless media) for the purpose of reducing any interferences or in order to avoid collisions that might occur. Our scenarios are based on the MAC layer enhancements in 802.11n and DCF based CSMA/CA (see Fig. 1 for more details). Firstly, we created a model for the two pairs scenario, shown in Fig. 2a. Then, we considered the three pairs scenario (see Fig. 2b) for comparison purposes.

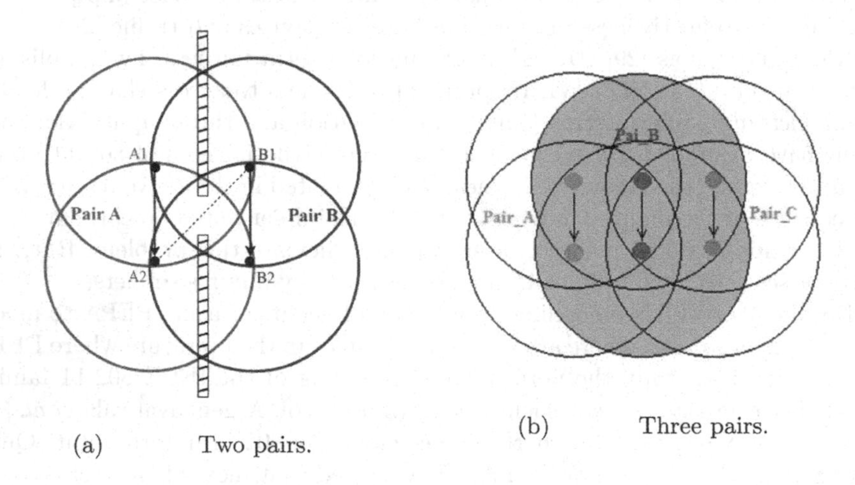

Fig. 2. The two pairs and three pairs scenarios.

5.1 The Two Pairs Scenario (Scenario 1)

We created a model for two pairs scenario by considering the row 5 in MAC layer enhancements (see Fig. 1). This row is useful to illustrate the behaviour of the transmitting pair, to provide a baseline performance. In this scenario, we focused on transferring block frames without repeated waiting time to send another frame, and with a single *BA* to acknowledge the entire frame burst. The purpose of this scenario is to investigate how the performance will be affected by the frame bursting when there is a competition without interference.

This scenario consists of *Pair_A* and *Pair_B*, which they are able to interact due to a wireless medium (*Medium_F*). Both pairs are symmetric, independent and may listen to the medium simultaneously before sending any frames. If the medium is not free, then the node waits for a period of time until the medium becomes idle again. Once the medium becomes idle, a node may start to transmit a sequence of frames. After each frame, a probabilistic choice is made to send a subsequent frame. Once the frame bursting is delivered successfully, a *BA* will acknowledge all the frames that have been sent. This scenario is fair in terms of channel access, as both pairs can hear each other and they can occupy the medium equally, but under the condition that both pairs are synchronised and have the same probability of sending successive frames.

We examined two case studies in this scenario by observing the number of frame bursting sent by both pairs. In the first case study, both pairs are symmetric and can use the medium equally, in which average frame burst length n where $2 \leq n \leq 10$ for both pairs are similar. The probability of subsequent frames in *Pair_A* is denoted by p_1 and in *Pair_B* denoted by p_2 i. e. $p_1{=}p_2$. However, in the second case study we show the effect of frame bursting in details, which p_1 in *Pair_A* is similar to p_1 in the first case study, but *Pair_B* has a fixed number of average frame burst length (p_2), i. e. $p_1 \neq p_2$.

In this model, `Pair_A` attempts to transmit after sensing the medium at the beginning to implement the transferring. `Pair_A` draws backoff to `Pair_A0`, then `Pair_A0` counts *DIFS* to `Pair_A1` or waits in `queue` as `Pair_A5`. A node waits at rate μ_{data} in `Pair_A5` before reaches to `Pair_A4`. `Pair_A4` has choices, either counts *DIFS* to `Pair_A1` or *EIFS* to `Pair_A1` or waits in the `queue` at `Pair_A5`. `Pair_A1` starts count `backoff` to stay at `Pair_A1` or ends `backoff` to `Pair_A2` or stays in the `queue` to `Pair_A5`. If it selects `Pair_A2`, then it starts to transmit a frame with a probability of p_1 to send a subsequent frame (consequently staying in `Pair_A2`) or it ends the transmission, by going to `Pair_A3` with a probability of (1-p_1). If any node occupies the medium then the other node may `queue` at `Pair_A5`. When it stops transmitting, then in `Pair_A3` *SIFS* counts to `Pair_A6`. Finally, `Pair_A6` sends a *BA* to acknowledge that all the fames received successfully. This process will apply to the second pair (`Pair_B`).

This system will be fair if $p_1{=}p_2$, i. e. that the frame burst lengths at each pair are identically distributed. In our results we will also explore the case study, where $p_1 \neq p_2$. The model of the two pairs scenario shows as follows.

$$
\begin{aligned}
Pair\_A &\stackrel{def}{=} (draw\_backoff, r).Pair\_A0\\
Pair\_A0 &\stackrel{def}{=} (count\_difsA, \mu_{difs}).Pair\_A1 + (queueA, \top).Pair\_A5\\
Pair\_A1 &\stackrel{def}{=} (count\_backoffA, p\mu_{bck}).Pair\_A1 + (queueA, \top).Pair\_A5\\
&+ (end\_backoffA, q\mu_{bck}).Pair\_A2\\
Pair\_A2 &\stackrel{def}{=} (transmitA, p_1\mu_{data}).Pair\_A2 + (queueA, \top).Pair\_A5\\
&+ (transmitA, (1-p_1)\mu_{data}).Pair\_A3\\
Pair\_A3 &\stackrel{def}{=} (count\_sifs, \mu_{sifs}).Pair\_A6\\
Pair\_A4 &\stackrel{def}{=} (count\_difsA, \mu_{difs}).Pair\_A1 + (count\_eifsA, \mu_{eifs}).Pair\_A1\\
&+ (queueA, \top).Pair\_A5\\
Pair\_A5 &\stackrel{def}{=} (wait, \mu_{data}).Pair\_A4\\
Pair\_A6 &\stackrel{def}{=} (BA\_A, \mu_{ack}).Pair\_A\\
Pair\_B &\stackrel{def}{=} (draw\_backoff, r).Pair\_B0\\
Pair\_B0 &\stackrel{def}{=} (count\_difsB, \mu_{difs}).Pair\_B1 + (queueB, \top).Pair\_B5\\
Pair\_B1 &\stackrel{def}{=} (count\_backoffB, p\mu_{bck}).Pair\_B1 + (queueB, \top).Pair\_B5\\
&+ (end\_backoffB, q\mu_{bck}).Pair\_B2\\
Pair\_B2 &\stackrel{def}{=} (transmitB, p_2\mu_{data}).Pair\_B2 + (queueB, \top).Pair\_B5\\
&+ (transmitB, (1-p_2)\mu_{data}).Pair\_B3\\
Pair\_B3 &\stackrel{def}{=} (count\_sifs, \mu_{sifs}).Pair\_B6\\
Pair\_B4 &\stackrel{def}{=} (count\_difsB, \mu_{difs}).Pair\_B1 + (count\_eifsB, \mu_{eifs}).Pair\_B1\\
&+ (queueB, \top).Pair\_B5\\
Pair\_B5 &\stackrel{def}{=} (wait, \mu_{data}).Pair\_B4\\
Pair\_B6 &\stackrel{def}{=} (BA\_B, \mu_{ack}).Pair\_B
\end{aligned}
$$

The Medium_F component shows the situations, where the medium is unoccupied, but `Medium_F1` presents the medium being used by the `Pair_B` and `Medium_F2` shows the medium being used by the `Pair_A`. The medium component between both pairs is presented as follows:

$$
\begin{aligned}
Medium\_F &\stackrel{def}{=} (transmitA, \top).Medium\_F2\\
&+ (transmitB, \top).Medium\_F1\\
&+ (count\_difsA, \top).Medium\_F\\
&+ (count\_eifsA, \top).Medium\_F\\
&+ (count\_backoffA, \top).Medium\_F\\
&+ (end\_backoffA, \top).Medium\_F\\
&+ (count\_difsB, \top).Medium\_F\\
&+ (count\_backoffB, \top).Medium\_F\\
&+ (end\_backoffB, \top).Medium\_F\\
&+ (count\_eifsB, \top).Medium\_F\\
Medium\_F1 &\stackrel{def}{=} (transmitB, \top).Medium\_F1 + (BA\_B, \top).Medium\_F\\
&+ (queueA, \lambda oc).Medium\_F1\\
Medium\_F2 &\stackrel{def}{=} (transmitA, \top).Medium\_F2 + (BA\_A, \top).Medium\_F\\
&+ (queueB, \lambda oc).Medium\_F2
\end{aligned}
$$

The Complete System: In this model, the `Pair_A`, `Pair_B` and `Medium_F` are interacting through the following cooperation sets:

$$Scenario1 \stackrel{def}{=} (Pair\_A \bowtie_K Medium\_F) \bowtie_L Pair\_B$$

where the values of K and L are:

$$K = \{\, transmitA, BA\_A, queueA, count\_difsA, count\_backoffA, end\_backoffA, count\_eifsA \}$$
$$L = \{\, transmitB, BA\_B, queueB, count\_difsB, count\_backoffB, end\_backoffB, count\_eifsB \}$$

5.2 The Three Pairs Scenario (Scenario 2)

In this section, we present three pairs scenario to study topographic unfairness in 802.11n protocol. Once again, we illustrated a model which a central pair in competition with the two external pairs (Fig. 2b). As in the two pair scenario, we modelled the most optimistic frame bursting option (row 5 in Fig. 1). The following specifies the behaviour of the three pairs (`Pair_A`, `Pair_B` and `Pair_C`) and the medium (`Medium_F`). We assume that the external pairs (`Pair_A` and `Pair_C`) have the same frame burst probability (p_1), but that may be different for the central pair (`Pair_B`) as we have denoted by p_2.

$$
\begin{aligned}
Pair\_A &\stackrel{def}{=} (draw\_backoff, r).Pair\_A0\\
Pair\_A0 &\stackrel{def}{=} (count\_difsA, \mu difs).Pair\_A1 + (queueA, \top).Pair\_A5\\
Pair\_A1 &\stackrel{def}{=} (count\_backoffA, p\mu_{bck}).Pair\_A1 + (queueA, \top).Pair\_A5\\
&\quad + (end\_backoffA, q\mu_{bck}).Pair\_A2\\
Pair\_A2 &\stackrel{def}{=} (transmitA, p_1\mu_{data}).Pair\_A2 + (queueA, \top).Pair\_A5\\
&\quad + (transmitA, (1-p_1)\mu_{data}).Pair\_A3\\
Pair\_A3 &\stackrel{def}{=} (count\_sifs, \mu_{sifs}).Pair\_A6\\
Pair\_A4 &\stackrel{def}{=} (count\_difsA, \mu_{difs}).Pair\_A1 + (count\_eifsA, \mu_{eifs}).Pair\_A1\\
&\quad + (queueA, \top).Pair\_A5\\
Pair\_A5 &\stackrel{def}{=} (wait, \mu_{data}).Pair\_A4\\
Pair\_A6 &\stackrel{def}{=} (BA\_A, \mu_{ack}).Pair\_A
\end{aligned}
$$

$$
\begin{aligned}
Pair\_B &\stackrel{def}{=} (draw\_backoff, r).Pair\_B0\\
Pair\_B0 &\stackrel{def}{=} (count\_difsB, \mu_{difs}).Pair\_B1 + (queueB, \top).Pair\_B5\\
Pair\_B1 &\stackrel{def}{=} (count\_backoffB, p\mu_{bck}).Pair\_B1 + (queueB, \top).Pair\_B5\\
&\quad + (end\_backoffB, q\mu_{bck}).Pair\_B2\\
Pair\_B2 &\stackrel{def}{=} (transmitB, p_2\mu_{data}).Pair\_B2 + (queueB, \top).Pair\_B5\\
&\quad + (transmitB, (1-p_2)\mu_{data}).Pair\_B3\\
Pair\_B3 &\stackrel{def}{=} (count\_sifs, \mu_{sifs}).Pair\_B6\\
Pair\_B4 &\stackrel{def}{=} (count\_difsB, \mu_{difs}).Pair\_B1 + (count\_eifsB, \mu_{eifs}).Pair\_B1\\
&\quad + (queueB, \top).Pair\_B5\\
Pair\_B5 &\stackrel{def}{=} (wait, \mu_{data}).Pair\_B4\\
Pair\_B6 &\stackrel{def}{=} (BA\_B, \mu_{ack}).Pair\_B
\end{aligned}
$$

$$
\begin{aligned}
Pair\_C &\stackrel{def}{=} (draw\_backoff, r).Pair\_C0\\
Pair\_C0 &\stackrel{def}{=} (count\_difsC, \mu_{difs}).Pair\_C1 + (queueC, \top).Pair\_C5\\
Pair\_C1 &\stackrel{def}{=} (count\_backoffC, p\mu_{bck}).Pair\_C1 + (queueC, \top).Pair\_C5\\
&\quad + (end\_backoffC, q\mu_{bck}).Pair\_C2\\
Pair\_C2 &\stackrel{def}{=} (transmitC, p_1\mu_{data}).Pair\_C2 + (queueC, \top).Pair\_C5\\
&\quad + (transmitC, (1-p_1)\mu_{data}).Pair\_C3\\
Pair\_C3 &\stackrel{def}{=} (count\_sifs, \mu_{sifs}).Pair\_C6\\
Pair\_C4 &\stackrel{def}{=} (count\_difsC, \mu_{difs}).Pair\_C1 + (count\_eifsC, \mu_{eifs}).Pair\_C1\\
&\quad + (queueC, \top).Pair\_C5\\
Pair\_C5 &\stackrel{def}{=} (wait, \mu_{data}).Pair\_C4\\
Pair\_C6 &\stackrel{def}{=} (BA\_C, \mu_{ack}).Pair\_C
\end{aligned}
$$

Component of Medium F: The shared medium component in the three pairs scenario shows, that the `Medium_F` represents the situation where the medium is unoccupied. `Medium_F1` represents the medium being used by the central pair (Pair_B).`Medium_F2` represents the medium being used by the external pair (Pair_A). Also, `Medium_F3` represents the medium being used by the external pair (Pair_C). Finally, `Medium_F4` represents the medium being used by both external pairs (Pair_A and Pair_C). The availability of all pairs to transmit interacts with the shared actions of the medium component as follows:

$$
\begin{aligned}
Medium\_F &\stackrel{def}{=} (transmitC, \top).Medium\_F3\\
&+ (transmitA, \top).Medium\_F2\\
&+ (transmitB, \top).Medium\_F1\\
&+ (count\_difsC, \top).Medium\_F\\
&+ (count\_backoffC, \top).Medium\_F\\
&+ (end\_backoffC, \top).Medium\_F\\
&+ (count\_eifsC, \top).Medium\_F\\
&+ (count\_difsA, \top).Medium\_F\\
&+ (count\_backoffA, \top).Medium\_F\\
&+ (end\_backoffA, \top).Medium\_F\\
&+ (count\_eifsA, \top).Medium\_F\\
&+ (count\_difsB, \top).Medium\_F\\
&+ (count\_backoffB, \top).Medium\_F\\
&+ (end\_backoffB, \top).Medium\_F\\
&+ (count\_eifsB, \top).Medium\_F\\
Medium\_F1 &\stackrel{def}{=} (transmitB, \top).Medium\_F1 + (BA\_B, \top).Medium\_F\\
&+ (queueA, \lambda oc).Medium\_F1 + (queueC, \lambda oc).Medium\_F1
\end{aligned}
$$

$$
\begin{aligned}
Medium\_F2 &\stackrel{def}{=} (transmitA, \top).Medium\_F2\\
&+ (transmitC, \top).Medium\_F4\\
&+ (BA\_A, \top).Medium\_F + (count\_difsC, \top).Medium\_F2\\
&+ (count\_backoffC, \top).Medium\_F2\\
&+ (end\_backoffC, \top).Medium\_F2\\
&+ (count\_eifsC, \top).Medium\_F2\\
&+ (queueB, \lambda oc).Medium\_F2\\
Medium\_F3 &\stackrel{def}{=} (transmitC, \top).Medium\_F3\\
&+ (transmitA, \top).Medium\_F4\\
&+ (BA\_C, \top).Medium\_F + (count\_difsC, \top).Medium\_F3\\
&+ (count\_backoffC, \top).Medium\_F3\\
&+ (end\_backoffC, \top).Medium\_F3\\
&+ (count\_eifsC, \top).Medium\_F3\\
&+ (queueB, \lambda oc).Medium\_F3\\
Medium\_F4 &\stackrel{def}{=} (BA\_A, \top).Medium\_F3 + (BA\_C, \top).Medium\_F2\\
&+ (transmitA, \top).Medium\_F4\\
&+ (transmitC, \top).Medium\_F4\\
&+ (queueB, \lambda oc).Medium\_F4
\end{aligned}
$$

The Complete System: In this model all components interact through this cooperation sets:

$$Scenario2 \stackrel{def}{=} ((Pair\_A \| Pair\_C) \underset{K}{\bowtie} Medium\_F) \underset{L}{\bowtie} Pair\_B$$

where the values of K and L are:

$$\begin{aligned} K = \{ & transmitA, BA\_A, queueA, count\_difsA, count\_backoffA, \\ & end\_backoffA, count\_eifsA, transmitC, BA\_C, queueC, \\ & count\_difsC, count\_backoffC, end\_backoffC, count\_eifsC\}. \\ L = \{ & transmitB, BA\_B, queueB, count\_difsB, count\_backoffB, \\ & end\_backoffB, count\_eifsB\}. \end{aligned}$$

6 Parameters

This section shows the main parameters in the IEEE 802.11n protocol. The IEEE 802.11 protocols have very specific inter-frame spacing, as they coordinate access to the medium for transmitting frames. When any pair wants to transmit, firstly it senses the channel to be used if it is idle, once silence is detected, then the node transmits with the probability of 'p'. For convenience, in our scenarios each pair has count back-off and end back-off actions with ($p\times$ μ_{bck}) and ($q\times$ μ_{bck}) rates respectively. We have assumed p and q=0.5 (where, q=1-p). According to the definition of 802.11n, the possible data rate per stream are 15, 30, 45, 60, 90, 120, 135 and 150 Mbit/s [33]. In this study, we considered 150 Mbit/s as a sample of data rates. This data rate applied with each of the frame payload size 700, 900, 1000, 1200, 1400 and 1500 bytes. The frames per time unit for arrival and departure rates are $\lambda oc = 100000$ and $\mu = 200000$ respectively. In our model μ_{ack} shows as a rate of ACK. If $\mu_{ack} = 1644.75$ for 1 Mbit/s then for 150 Mbit/s speed the $\mu_{ack} = 246712.5$ .5 (150×1644.75).

$$\mu_{ack} = \frac{\text{channel throughput}}{ACK \text{ length}}$$

where the value of ACK length = 1 byte.

Inter-Frame Space (*IFS*): A small amount of time is required in the IEEE 802.11 families to generate a successful transmission for interface protocol. At the beginning, the length of the *IFS* is dependent on the previous frame type, if any noise occurs, the *IFS* is used. Possibly, if transmission of a particular frame ends and before another one starts, the *IFS* applies a delay to the channel to stay clear. It is an essential idle period of time needed to ensure that other nodes may access the channel. The *IFS* is to supply a waiting time for each frame transmission in a specific node, allowing the transmitted signal to reach another node (essential for listening) that it is measured in microseconds [34–36].

Short Inter-Frame Space (*SIFS*): *SIFS* is the shortest IFS for highest priority transmissions used with DCF, that it aims to process a received frame;

for instance, *SIFS* is applied between a data frame and the *ACK*. Additionally, *SIFS* can be used to separate single frames in a back-to-back frame burst, and the value of *SIFS* time in IEEE 802.11n protocol is 16 μs.

Slot Time : It is an integral required number if node attempts to send a data with the beginning of the transmission slot boundary. The duration of slot time is designed to provide sufficient space of the variability and sufficient required time to transmit a node's preamble to be detected by other nodes before the next slot boundary. The slot time in the 802.11n is 9 μs (for 5 GHz).

DCF Inter-Frame Space (*DIFS*): *DIFS* is a medium priority waiting time used by nodes, that operates under the DCF to send and manage the data frames. The *DIFS* can be used to monitor the medium for a longer period of time than *SIFS*. If the channel is idle, the node waits for the *DIFS* to determine that the channel is not being used, and it waits for another period of time (*backoff*). The following formula shows the definition of *DIFS*.
DIFS = *SIFS* + (2 × (slot time = 9 μs in IEEE 802.11n standard)).

Extended Inter-Frame Space (*EIFS*): If the node detects a signal and a frame is not correctly received, the transmission node uses *EIFS* instead of *DIFS* (used with erroneous frame transmission) while an *ACK* might not be detected. *EIFS* is the longest of *IFS*, but it has the lowest priority after *DIFS*.
EIFS = *SIFS* + *DIFS* + transmission time (*ACK*-lowest basic rate). Where lowest basic rate *ACK* is the time required to transmit an *ACK* frame at the lowest mandatory PHY data rate. The *EIFS* in IEEE 802.11n devices using OFDM is 160 μs.

Contention Window (*CW*): In CSMA/CA, if a node tries to send any frame, firstly it senses the channel. The node transmits if it is free; if not, the node waits for a random backoff, and an integer value selected by the node from a (*CW*), until it becomes free. If the states of medium transition is changed from busy to free, multi nodes might attempt to occupy the medium for sending data. In this case, collision might occur in transmission, and to minimise this negative impact on the medium, the node waits with a random backoff count and defer for that number of slot times. This is the main intention to minimise any collision once it experiences an idle channel for an appropriate *IFS*, otherwise many waiting nodes might transmit simultaneously. The node needs less time to wait if there is a shorter backoff period, so transmission will be faster, unless there is a collision. The random backoff count is selected as an integer drawn from uniform distribution that is chosen in [0, *CW*]. *CW* = *CWmin* for all nodes if a node successfully transmits a packet and receives an *ACK*. Otherwise, the node draws another *backoff* and the *CW* increases exponentially, until it reaches *CW* = *CWmax*. Finally, when the backoff reaches 0, the node starts to transmit and *CW* resets to the initial value of *CW* = *CWmin* when the frame is received. In 802.11n, if *CWmin* = 15 then *CWmax* = 1023 by augmented the *CWmin* to 2n-1 on each retry. Backoff Time = (Random() mod (*CW*+1)) × slot time.

If BackoffTimer = b, where b is a random integer, also $CWmin \leq b \leq CWmax$. By using the mean of $CWmin$ and if slot time = 9 μs, then we can calculate μbck:

$$\mu_{bck} = \frac{10^6}{\bar{CW} \times \text{slot time}}$$

Accordingly, if the mean of $CWmin = 7.5$ then $\mu bck = 14814.81481$. Additionally, if p = 0.5 then $p\mu bck = 7407.407407$. Likewise, when $DIFS = 34$ and $SIFS = 16$, then we can find the $\mu difs$ and $\mu sifs$ as follow:

$$\mu_{difs} = \frac{10^6}{34}, \text{ and } \mu_{eifs} = \frac{10^6}{16}$$

Finally, the receiver will send an ACK if it obtains a frame successfully. Similarly, $\mu data$ can be found as shows in the following. For example, the $\mu data$ is equal to 26785.71428 for 700 packet payload size in 150 Mbit/s data rate:

$$\mu_{data} = \frac{\text{data rate} \times \frac{10^6}{8}}{\text{packet payload size}}$$

Probability of Subsequent Frames p_1 and p_2: In our study, we have observed the number of frame bursting to be sent by pair(s). Initially, the average frame burst length is denoted by n, where $2 \leq n \leq 10$ for one pair and two pairs scenario, however $3 \leq n \leq 10$ for three pairs scenario. Also, the probability of subsequent frames to be sent is denoted by p_1, where $0 \leq p_1 \leq 1$. Thus, we can do the following to calculate p_1 and n, where $n \in \{1, 2, 3, 4, 5, 7, 8, 9, 10\}$:

$$n = \frac{1}{1 - p_1}, \text{ hence } p_1 = \frac{(n-1)}{n}$$

In our study, we demonstrated analytically the effect of frame bursting, that a pair(s) has a variation of frame bursting to be sent per time interval, in a specific pair is denoted by p_1. In our scenario(s), if two or more pairs have the same behaviour, then they all have the same probability of subsequent frames. But, if any pair has no variation of frame bursting, then the probability of subsequent frames is fixed, which can be shown as p_2. In two pairs case, if both pairs are asymmetric then $Pair\_A$ has p_1 and $Pair\_B$ has p_2 (where $n = 2$ then $p_2 = 0.5$) i. e. $p_1 \neq p_2$, but in the three pairs case p_2 for the central pair ($Pair\_B$) is equal to 0.66, where $n = 3$. Further details are shown in next section.

Table 1. Parameter values of IEEE 802.11n.

Attribut	Typical vale in 802.11n
CWmin, and CWmax	15 (mean = 7.5), and 1023
Slot time	20{µs} or 9{µs} ms
SIFS	16{µs} ms
DIFS	34{µs} ms
EIFS	160{µs} ms
p , q (q = 1-p)	0.5
λoc	100000
μ	200000
μack (for 150 Mbps)	246712.5
$\mu data$ (for 700 packet)	26785.71428
μbck (if CW mean = 7.5)	14814.81481
pμbck (if p = 0.5)	7407.407407

7 Results and Discussions

This section shows the obtained results of our experimental study on the IEEE 802.11n, as we considered 150 Mbit/s data rates. The specification in our PEPA models with the presented parameters in the previous sections are used to measure the channel utilisation rate and channel throughput, to analyse the performance fairness of this protocol. This section will present the results of the two pairs and three pairs scenarios in terms of channel access.

7.1 Performance Results of the Two Pairs Scenario (Scenario 1)

We studied the effect of average frame bursting length for the two pairs scenario in two case studies. Firstly, we investigated both pairs that have the same number of frame bursting, where p_1=p_2. Then, we analysed the model if a particular pair has several frame bursting to be sent during transmission, but `Pair_B` has a fixed number of frame bursting to be sent, i. e. $p_1 \neq p_2$.

Figure 3 shows the utilisation for a particular pair (`Pair_A`) in the two pairs scenario, where both pairs are symmetric and can send the same block of frames in transmission, i. e. p_1=p_2. The utilisation in `Pair_A` increases as the packet payload size increases. When the node sends a large number of frame bursting, it obtains the highest utilisation, therefore the node will occupy the medium for longer. The utilisation in the second pair (`Pair_B`) is similar to the utilisation in the first pair (`Pair_A`) as they are symmetric and sharing the medium equally.

Figure 4 shows that the channel throughput rate decreases for a particular pair (such as `Pair_A`) as the packet payload size increases. In this scenario, if the medium is free to be used then both pairs have the same chance to occupy it, as they have the same number of frames to be sent, where p_1=p_2. Also, the

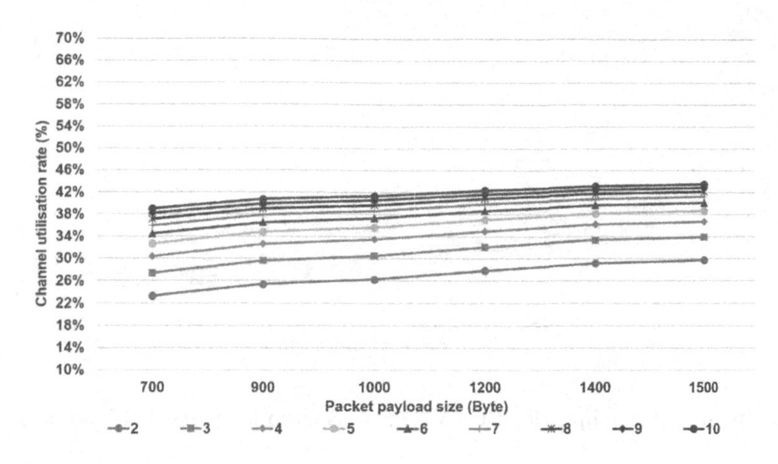

Fig. 3. Channel utilisation for the two pairs scenario for pair A/B, where $p_1=p_2$.

channel throughput decreases in each node if they send larger packets with a larger number of frame bursting. Hence, less time is needed to transmit smaller packet payload for smaller block of frames (it is similar in `Pair_B`).

In the second case study we analysed the utilisation, where $p_1 \neq p_2$. Figure 5 shows that the utilisation in the first pair (`Pair_A`) increases as the packet payload size increases, while this pair sends several frame bursting by increasing the probability of sequence frame bursting, i. e. the medium will be occupied longer by this pair as the average frame burst length n increases from 2 to 10.

Figure 6 shows that the utilisation in `Pair_B` increases as the packet payload size increases, where this pair has a fixed frame bursting $n = 2$, but in `Pair_A` n increases from 2 to 10. However, if we compare the utilisation `Pair_B` to the utilisation in `Pair_A`, we can understand that the utilisation in `Pair_B` is significantly lower than the utilisation in the `Pair_A` as the number of frame bursting to be sent is increased in `Pair_A`. When the average frame burst length n increases from 2 to 10 in `Pair_A`, `Pair_B` will be completely prevented from transmitting, because the `Pair_A` can send larger frame bursting and uses the medium longer, while `Pair_B` has less chance to use the medium. This increase of the average frame burst length n in `Pair_A` affect the utilisation in `Pair_B`, that decreases the channel utilisation in `Pair_B` rather than increasing it.

In second case study, the transmission of larger frames in `Pair_A` and fixed frames in `Pair_B` have pros and cons. The channel throughput decreases in both pairs as the package payload size increases, see Figure `Pair_A` in Fig. 7. But, when the second pair (`Pair_B`) has no variation of frame bursting to be sent (it sends two frames in per circle of transmission) then the channel throughput decreases more when the average frame burst length n is larger in `Pair_A`, as shown in Fig. 8. In this case, `Pair_A` will stay longer in the medium and `Pair_B` has less chance to access the medium (`Pair_A` will send frames for much longer than `Pair_B`, and `Pair_B` waits for much longer in the queue than `Pair_A`). `Pair_A` will gain more advantages as it transmits more frames than `Pair_B`. Hence the

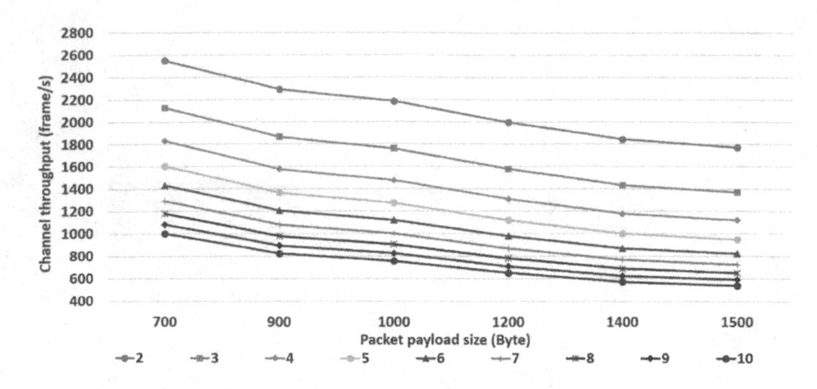

Fig. 4. Channel throughput for the two pairs scenario for pairs A/B, where $p_1 = p_2$.

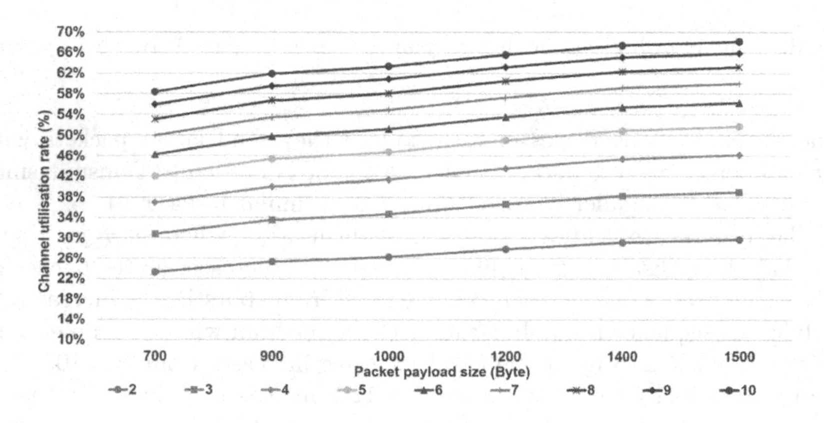

Fig. 5. Channel utilisation for the two pairs scenario for pair A, where $p_1 \neq p_2$.

reduction in Pair_B has affected the efficient usage of the medium in terms of channel access and sending block frames approaches.

7.2 Performance Results of the Three Pairs Scenario (Scenario 2)

This section shows the results of the three pairs scenario. We considered this scenario for the purpose of reducing the unfairness in a particular pair (the inner pair) and investigate the outer pairs, while many unfairness has been reported in ([4,37–39]) regards to IEEE 802.11b and 11g. We modelled the two case studies in this scenario as shown in Sect. 5.2. Due to these case studies, we investigated how the performance of 802.11n changes by concentrating on the frame bursting to improve the efficiency of the protocol and channel throughput. The results of each case shows as follows.

All pairs in the three pairs scenario compete to use the medium. If the medium is free to be used, then any pair can attempt to transmit. If the inner pair uses the medium, then both outer pairs will wait until the medium becomes

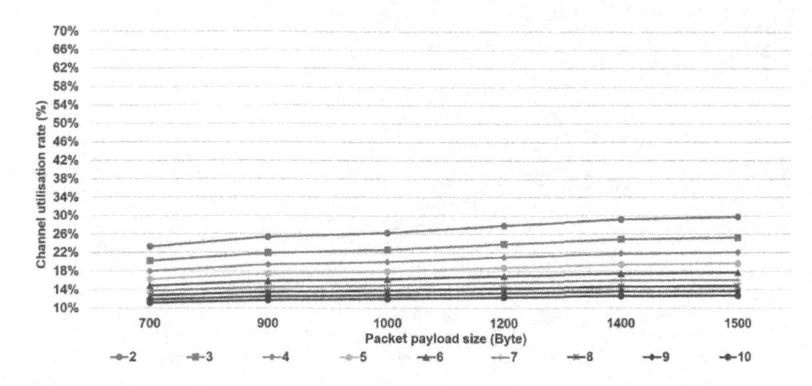

Fig. 6. Channel utilisation for the two pairs scenario for pair B, where $p_1 \neq p_2$.

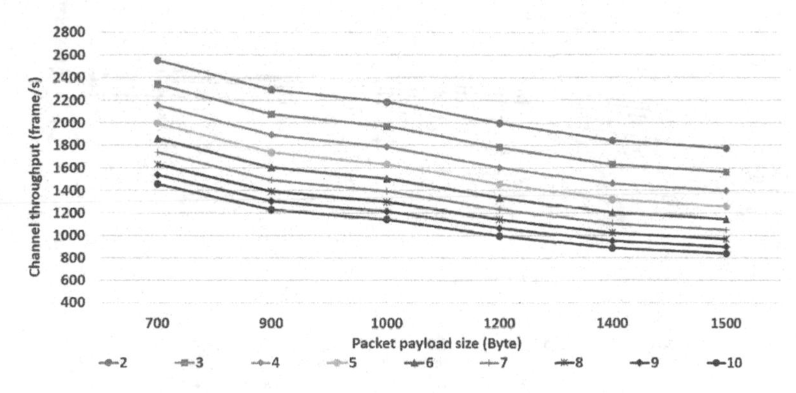

Fig. 7. Channel throughput rate in the two pairs scenario for pair A, where $p_1 \neq p_2$.

idle. But, if the medium is being in use by one of the outer pairs, only the other outer pair can transmit and the inner pair will penalised.

In the first case study in this scenario, once all pairs have the same number of frame bursting to be sent at the same time, where $p_1 = p_2$, then the channel utilisation of outer pairs increases as the packet payload size increases. Because both outer pairs cannot hear each other, they have a higher chance to use the medium to send more frames. Hence, the channel utilisation in the outer pairs will be higher as the average length of frame bursting n becomes higher, in this scenario the value of n assumed to be 3 as an initial value, see Fig. 9.

The utilisation in the inner pair is different to the utilisation in outer pairs. In the first case study, as the average frame burst length n increases from 3 to 10 and the package payload increases too, the inner pair will have less opportunity to access the medium. In this circumstance the utilisation in the inner pair is completely congested. Hence, when the frame intensity becomes less (the average frame burst length n decreases) then there is more opportunity for the inner pair to access the medium. In this case the inner pair might be affected less and the proportional of using the medium will be greater, see Fig. 10.

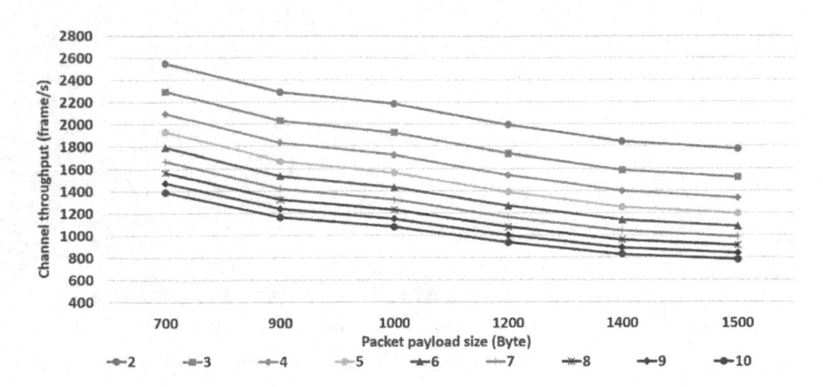

Fig. 8. Channel throughput rate in the two pairs scenario for pair B, where $p_1 \neq p_2$.

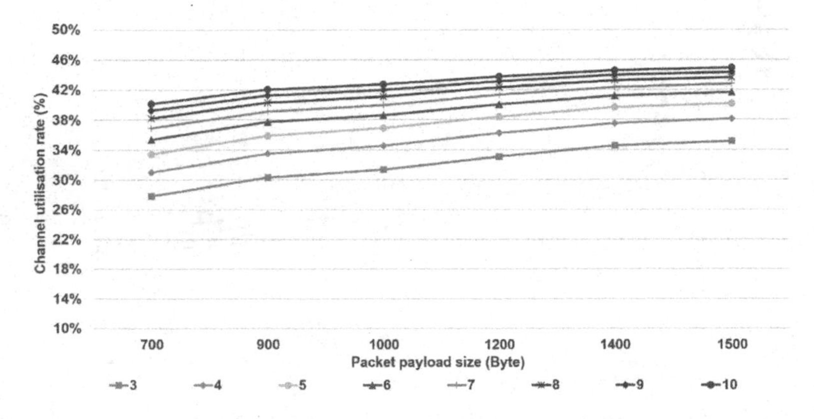

Fig. 9. Channel utilisation for the outer pairs in the three pair scenario, where $p_1{=}p_2$.

Likewise, in our first case study in this scenario, we also examined the throughput. It decreases in all pairs as the packet payload size increases, because of the channel occupancy time. The outer pairs have higher throughput relative to packet payload size with the number of frame bursting. When any outer pairs are sending larger packets with a larger average frame burst length n, the throughput decreases as it sends more and sends larger frame bursting, see Fig. 11. The smaller frames occupy less time in this channel. However, the throughput of the inner pair decreases more as the packet payload size increases compared to the throughput of the outer pairs, see Fig. 12. Hence, the inner pair can hear both outer pairs and will be blocked more the outer pairs. Therefore, both outer pairs have more chance to occupy the medium (with the longer sequence of frames in the inner pair, which we can understand that the worst throughput will obtain). This scenario is unfair in terms of channel throughput, as it is not significant for all pairs. Thus, the inner pair is out computed by both outer pairs and it is unfairly disadvantaged but the outer pairs are fairly advantaged. Hence, we have studied more details on the inner pair in our second

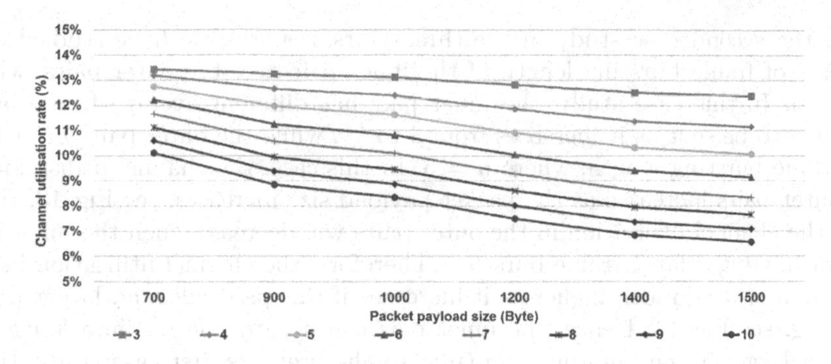

Fig. 10. Channel utilisation for the inner pair in the three pair scenario, where $p_1 = p_2$.

case study, as we have increased the number of frame bursting to be sent by the inner pair, in order to increase the fairness issue in this scenario.

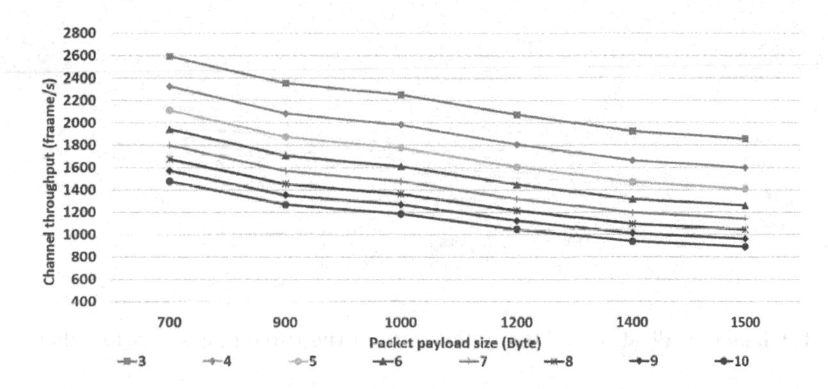

Fig. 11. Channel throughput for the outer pairs in the three pair scenario, where $p_1 = p_2$.

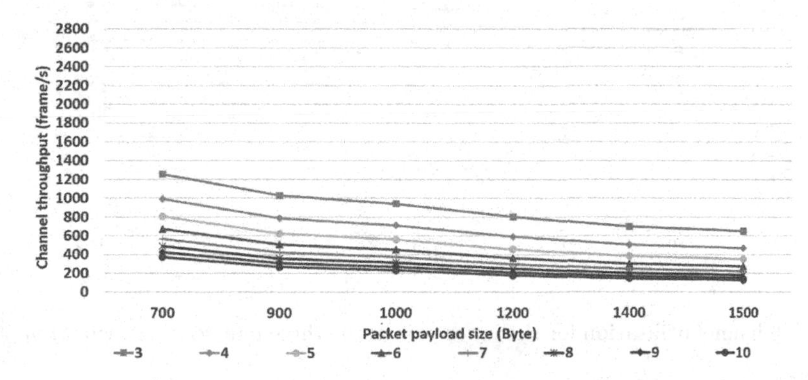

Fig. 12. Channel throughput for the inner pair in the three pair scenario, where $p_1 = p_2$.

In the second case study in the three pairs scenario, we have studied consistency of frame bursting length of the inner pair over the outer pairs, where $p_1 \neq p_2$. In this case study, the inner pair has different average frame burst length n to be sent as it increases from 3 to 10, while the outer pairs have only one frame bursting length, where $n = 3$. In this case, the channel utilisation in the outer pairs increases as the packet payload size increases, see Fig. 13. However, the channel utilisation in the outer pairs will decrease when the inner pair is transmitting a large frame bursting. Therefore, the channel utilisation in the inner pair will be much higher as it increases if this pair will send larger frame bursting, see Fig. 14. Hence, the inner pair can occupy the medium longer by sending larger frame bursting compared to the previous first case study. Here, we can understand that the channel utilisation will be higher if the inner pair is having more frame bursting and larger frame bursting length in transmission.

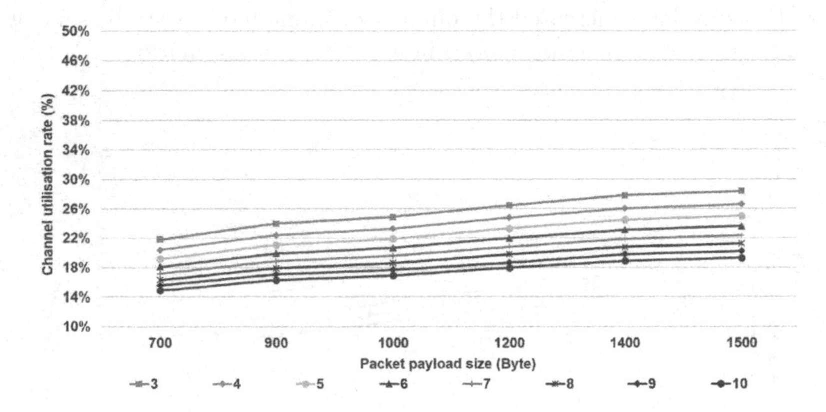

Fig. 13. Channel utilisation for the outer pairs in the three pair scenario, where $p_1 \neq p_2$.

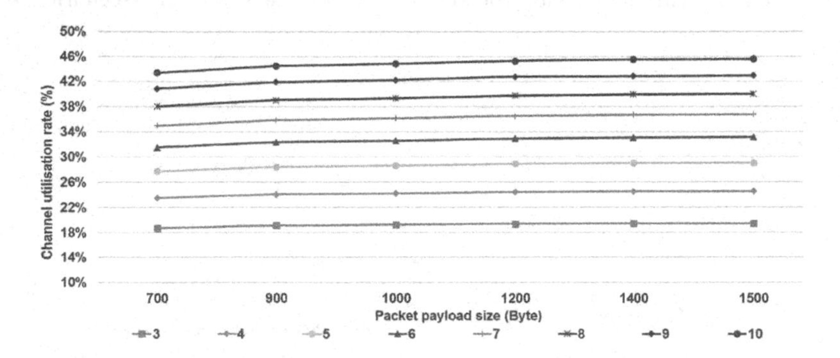

Fig. 14. Channel utilisation for the inner pair in the three pair scenario, where $p_1 \neq p_2$.

Finally, the throughput decreases as the packet payload size increases in all pairs in this case study. The throughput in the outer pairs and inner pair shows in Figs. 15 and 16 respectively. In this case, we observed that the throughput in the inner pair is much higher than the throughput in same pair in comparison to the previous case study. As the inner pair in our second case study has different frame bursting in transmission, that stays longer in transmitting. In terms of channel throughput and performance fairness, the second case addressed very low unfairness compared to the previous case. Thus, we can profitably gain less unfairness if the inner pair sends longer frame bursting in transmission.

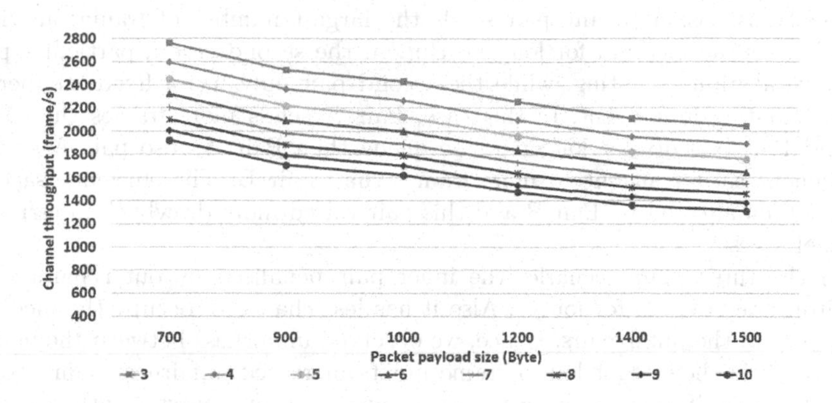

Fig. 15. Channel throughput for the outer pairs in the three pair scenario, where $p_1 \neq p_2$.

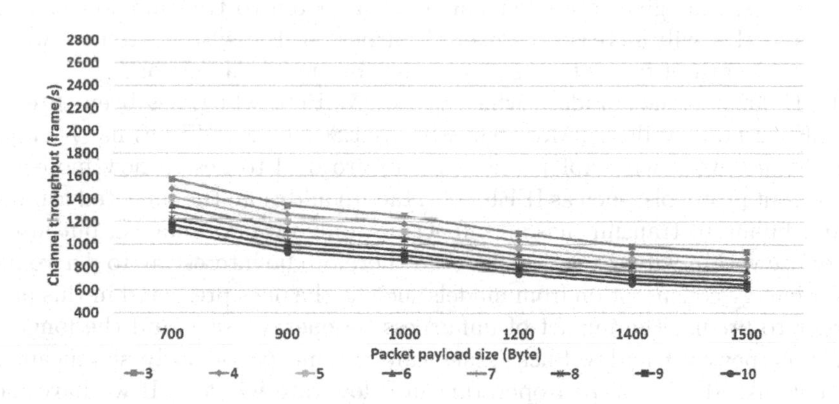

Fig. 16. Channel throughput for the inner pair in the three pair scenario, where $p_1 \neq p_2$.

8 Conclusion

This paper presented an analytical based model on IEEE 802.11n MAC layer enhancements used PEPA (over 150 Mbit/s bandwidth). We investigated the performance fairness between different communicating pairs of nodes within a restricted network topology. We illustrated the two pairs and three pairs scenarios focused on CSMA/CA for collision avoidance, performance improvement and accessing the medium; in terms of medium utilisation and throughput. As we analysed 802.11n when a pair sends multiple frames without repeated waiting and uses a single BA, to acknowledge the entire frames.

The two pairs scenario explored the effect of frame bursting length in two case studies. In the first case both pairs have the same number of frame bursting during transmission (if any pair sends the largest number of frame bursting, then it uses the medium for longer). But, in the second case a particular pair has several frame bursting, while the second pair only has a fixed number of frame bursting to be sent. In this case, Pair A sends more frames than Pair B, and Pair B waits for longer in the queue than Pair A, also pair A gained the benefit as it transmitted more frames than Pair B. The efficient usage of the medium affected by Pair B and this pair raised more drawback in terms of channel access.

In the three pairs scenario, the inner pair penalised by outer pairs as it monitors the medium for longer. Also it has less chance to occupy the medium compared to the outer pairs. Hence, we observed unfairness between the nodes. This scenario shows that longer frame bursts increased unfairness. Our results presented that if all pairs have the same average frame burst length n simultaneously, then the inner pair transmits less, while the outer pairs will occupy the channel for longer. This eventually limits the inner pair to transmit that increases the unfairness. However, the outer pairs can be restricted to send short frame bursts, that gives the inner pair more freedom to transmit longer frame bursts and this will increase the overall fairness of the system. Hence, we can utilise the length of frame bursts to promote fairness in a network.

IEEE 802.n is the standard which defines WiFi 4, which has been extended in IEEE802.11ac (WiFi 5). More recently IEEE802.11ax (WiFi 6) has been proposed Future work will explore the existing protocol to design new models for more recent protocols, such as IEEE 802.11ac. In addition, the repeated attempting and failing to transmit has an impact not only on performance, but also on energy use within wireless nodes. It would therefore be interesting to derive metrics for energy consumption from models such as the ones presented in this paper in order to predict the impact of unfairness on energy usage and the longevity of battery powered nodes. Such issues can become particularly significant for protocols like IEEE 802.15.4 operating in a low-rate wireless. If we have more congestion or collision between nodes, then more waiting time between nodes will be required, in these circumstances more energy is needed to resend data. As a consequence, by having the contention or competition between wireless nodes, then more power will be used by each node. Better understanding the relationship between throughput, fairness and energy usage in wireless networks is an area of significant theoretical challenge and practical interest.

Acknowledgements. This research is funded by Newcastle University and KRG Government.

References

1. Gast, M.: 802.11 wireless networks: the definitive guide. Southeast University Press (2006)
2. Pahlavan, K., Krishnamurthy, P.: Principles of wireless networks: a unified approach. Prentice Hall (2011)
3. Kim, J., Helmy, A.: The evolution of WLAN user mobility and its effect on prediction. In: 7th International Wireless Communications and Mobile Computing Conference (2011)
4. Othman Abdullah, C., Thomas, N.: Modelling unfairness in IEEE 802.11g networks with variable frame length. In: Wittevrongel, S., Phung-Duc, T. (eds.) ASMTA 2016. LNCS, vol. 9845, pp. 223–238. Springer, Cham (2015). https://doi.org/10.1007/978-3-319-43904-4_16
5. Friedrich, J., Frohn, S., Gubner, S., Lindemann, C.: Understanding IEEE 802.11n multi-hop communication in wireless networks. In: IEEE International Symposium on Modeling and Optimization in Mobile, Ad Hoc and Wireless Networks (WiOpt) (2011)
6. Alekhya, T., Mounika, B., Jyothi, E., Bhandari, B.: A waiting-time based backoff algorithm in the IEEE 802.11 based wireless networks. In: National Conference on Communications (2012)
7. Nandiraju, N., Gossain, H., Cavalcanti, D., Chowdhury, K., Agrawal, D.: Achieving fairness in wireless LANs by enhanced IEEE 802.11 DCF. In: IEEE International Conference on Wireless and Mobile Computing, Networking and Communications (2006)
8. Mohammad, N., Muhammad, S.: Modeling and analyzing MAC frame aggregation techniques in 802.11n using bi-dimensional Markovian model. In: Benlamri, R. (ed.) NDT 2012. CCIS, vol. 293, pp. 408–419. Springer, Heidelberg (2012). https://doi.org/10.1007/978-3-642-30507-8_35
9. Wang, C., Wei, H.: IEEE 802.11n MAC enhancement and performance evaluation. Mobile Netw. Appl. **14**(6), 760– 771 (2009). https://doi.org/10.1007/s11036-008-0129-2
10. Kolap, J., Krishnan, S., Shaha, N.: Frame aggregation mechanism for high throughput 802.11n WLANs. Int. J. Wirel. Mobile Netw. **4**(3), 141–153 (2012)
11. Hillston, J.: A compositional approach to performance modelling, vol. 12. Cambridge University Press (2005)
12. Tribastone, M., Duguid, A., Gilmore, S.: The PEPA Eclipse Plug-in. Perform. Eval. Rev. **36**(4), 6 (2009)
13. Singh, R.A., Indu, P.: Performance analysis of IEEE 802.11 in the presence of hidden terminal for wireless networks. In: Jain, L., Behera, H., Mandal, J., Mohapatra, D. (eds.) Computational Intelligence in Data Mining - Volume 1. Smart Innovation, Systems and Technologies, vol. 31. Springer, New Delhi (2015). https://doi.org/10.1007/978-81-322-2205-7_61
14. Pham, D., Sekercioglu, Y., Egan, G.: Performance of IEEE 802.11b wire915 less links: an experimental study. In: Proceedings of the IEEE Region 10 Conference (TENCON 2005) (2005)
15. Puigjaner, R.: Performance modelling of computer networks. In: Proceedings of the 2003 IFIP/ACM Latin America conference on Towards a Latin American Agenda for Network Research, no. 106–123 (2003)
16. Min, G., Wu, Y., Li, K., Al-Dubai, A.: Performance modelling and optimization of integrated wireless LANs and multi-hop mesh networks. Int. J. Commun. Syst. 23(9–10), 1111–1126 (2010)

17. Lee, H.: A MAC layer throughput over error-free and error-prone channel in the 802.11 a/g-based mobile LAN. In: Proceedings of the 9th Malaysia International Conference on Communications (2009)
18. Vučinič, M., Tourancheau, B., Duda, A.: Simulation of a backward compatible IEEE 802.11g network: access delay and throughput performance degradation. In: Mediterranean Conference on Embedded Computing (MECO) (2012)
19. Ho, M., Wang, J., Shelby, K., Haisch, H.: IEEE 802.11g OFDM WLAN throughput performance. In: Proceedings of the 58th IEEE Vehicular Technology Conference, vol. 4 (2004)
20. Khanduri, R., Rattan, S., Uniyal, A.: Understanding the features of IEEE 802.11g in high data rate wireless LANs. Int. J. Comput. Appl. **64**(8), 1–5 (2013)
21. Galloway, M.: Performance measurements of coexisting IEEE 802.11g/n networks. In: Proceedings of the 49th Annual Southeast Regional Conference. ACM (2011)
22. Fiehe, S., Riihijjarvi, J., Mahonen, P.: Experimental study on performance of IEEE 802.11n and impact of interferers on the 2.4 GHz ISM band. In: Proceedings of the 6th International Wireless Communications and Mobile Computing Conference (2010)
23. Xiao, Y.: IEEE 802.11n: enhancements for higher throughput in wireless LANs. IEEE Wirel. Commun. **12**(6), 82–91 (2005)
24. Hajlaoui, N., Jabri, I., Jemaa, M.: Experimental performance evaluation and frame aggregation enhancement in IEEE 802.11n WLANs. Int. J. Commun. Netw. Inf. Secur. **5**(1), 48–58 (2013)
25. Char, E., Chaari, L., Kamoun, L.: Fairness of the IEEE 802.11n aggregation scheme for real time application in unsaturated condition. In: 4th Joint IFIP Wireless and Mobile Networking Conference. IEEE (2011)
26. Mohamedou, A., Othman, M.: Analytical evaluation of unfairness problem in wireless LANs. arXiv preprint arXiv:1002.4833 (2010)
27. Zhai, H., Kwon, Y., Fang, Y.: Performance analysis of IEEE 802.11 MAC protocols in wireless LANs. Wirel. Commun. Mobile Comput. **4**(8), 917–931 (2004)
28. Shehadeh, Y., Chasaki, D.: Secure and efficient medium access in wireless networks.In: Proceedings of the 4th IEEE International Conference on Consumer Electronics (2014)
29. Kloul, L., Valois, F.: Investigating unfairness scenarios in MANET using 802.11b. In: Proceedings of the 2nd ACM International Workshop on Performance Evaluation of Wireless Ad hoc, sensor, and ubiquitous networks (2005)
30. Duda, A.: Understanding the performance of 802.11 networks. In: Proceedings of the 19th International Symposium on Personal, Indoor and Mobile Radio Communications, vol. 8 (2008)
31. Argent-Katwala, A., Bradley, J., Geisweiller, N., Gilmore, S., Thomas, N.: Modelling tools and techniques for the performance analysis of wireless protocols. In: Advances in Wireless Networks: Performance Modelling, Analysis and Enhancement, Nova Science Pub Inc. (2008)
32. Sridhar, K.N., Ciobanu, G.: Describing IEEE 802.11 wireless mechanisms by using the π-calculus and performance evaluation process algebra. In: Núñez, M., Maamar, Z., Pelayo, F.L., Pousttchi, K., Rubio, F. (eds.) FORTE 2004. LNCS, vol. 3236, pp. 233–247. Springer, Heidelberg (2004). https://doi.org/10.1007/978-3-540-30233-9_18
33. Ting, K., Ee, G., Ng, C., Noordin, N., Ali, B.: The performance evaluation of IEEE 802.11 against IEEE 802.15.4 with low transmission power. In: the 17th Asia Pacific Conference on Communications. IEEE (2011)

34. Ting, K., Kuo, F., Hwang, B., Wang, H., Lai, F.: An accurate power analysis model based on MAC layer for the DCF of 802.11n. In: International Symposium on Parallel and Distributed Processing with Applications, pp. 350–358 (2010)
35. Ginzburg, B., Kesselman, A.: Performance analysis of A-MPDU and AMSDU aggregation in IEEE 802.11n. In: IEEE Sarno Symposium (2007)
36. Perahiaa, E., Stacey, R.: Next generation wireless LANs: 802.11n and 802.11ac. Cambridge University Press (2013)
37. Abdullah, C., Thomas, N.: Formal performance modelling and analysis of IEEE 802.11 wireless LAN protocols. In: UK Performance Engineering Workshop, University of Leeds (2015)
38. Abdullah, C., Thomas, N.: Performance modelling of IEEE 802.11g wireless LAN protocols. In: 9th EAI International Conference on Performance Evaluation Methodologies and Tools (2015)
39. Abdullah, C.O., Thomas, N.: A PEPA model of IEEE 802.11b/g with hidden nodes. In: Fiems, D., Paolieri, M., Platis, A.N. (eds.) EPEW 2016. LNCS, vol. 9951, pp. 126–140. Springer, Cham (2016). https://doi.org/10.1007/978-3-319-46433-6_9

Author Index

M. Forshaw et al. (Eds.): PASM 2022, CCIS 1786, p. 133, 2023.
https://doi.org/10.1007/978-3-031-44053-3

Printed in the United States
by Baker & Taylor Publisher Services

Printed in the United States
by Baker & Taylor Publisher Services